Mathematicians Playing Games

Mathematicians Playing Games explores a wide variety of popular mathematical games, including their historical beginnings and the mathematical theories that underpin them. Its academic level is suitable for high school students and higher, but people of any age or level will find something to entertain them, and something new to learn. It offers a fantastic resource for high school mathematics classrooms or undergraduate mathematics for liberal arts course and belongs on the shelf of anyone with an interest in recreational mathematics.

Features:

* Suitable for anyone with an interest in games and mathematics, and could be especially useful to middle and high school students and their teachers

* Includes various exercises for fun for readers

Jon-Lark Kim was born in 1970 in South Korea. He received the BS degree in Mathematics from POSTECH, the MS degree in Mathematics from Seoul National University, South Korea, in 1997, and the PhD degree from the Department of Mathematics, University of Illinois at Chicago, in 2002. He was an Associate Professor at the University of Louisville until 2012. He is currently a Professor at the Department of Mathematics, Sogang University, and the Director of the Institute for Mathematical and Data Science, Seoul, South Korea. He is also CEO of DeepHelix, a start-up based on AI and healthcare.

He has authored more than 70 research papers on Coding Theory, Combinatorics, Cryptography, Games, and Machine Learning and a book titled *Selected Unsolved Problems in Coding Theory*. He is a Co-Editor of *Concise Encyclopedia of Coding Theory* published by Chapman and Hall/CRC in 2021. He was a recipient of the 2004 Kirkman Medal from the Institute of Combinatorics and its Applications.

He is a member of the Editorial Board of *Designs, Codes and Cryptography* and *Journal of Applied Mathematics and Computing*. His research interests include Coding Theory, Cryptography, Informatics, Fuzzy Theory, and Artificial Intelligence, hoping to find a method to unify all of these+alpha in his lifetime just as Rene Descartes discovered a Cartesian coordinate system which unifies Algebra and Geometry and included a quotation "Cogito ergo sum" ("I think, therefore I am") which unifies the human mind and body. His hobbies include playing soccer, reading, and writing.

AK Peters/CRC Recreational Mathematics Series

Series Editor: *Robert Fathauer, Snezana Lawrence, Jun Mitani, Colm Mulcahy, Peter Winkler, Carolyn Yackel, Luck, Logic, and White Lies*

For more information about this series please visit: https://www.routledge.com/
AK-PetersCRC-Recreational-Mathematics-Series/book-series/RECMATH

Mathematicians
Playing Games

Jon-Lark Kim
Sogang University, Korea

CRC Press
Taylor & Francis Group
Boca Raton London New York

CRC Press is an imprint of the
Taylor & Francis Group, an **informa** business

AN A K PETERS BOOK

Front Cover Image: veeravich/Shutterstock

First edition published 2024
by CRC Press
2385 NW Executive Center Drive, Suite 320, Boca Raton FL 33431

and by CRC Press
4 Park Square, Milton Park, Abingdon, Oxon, OX14 4RN

CRC Press is an imprint of Taylor & Francis Group, LLC

© 2024 Jon-Lark Kim

ISBN: 978-1-032-21361-3 (hbk)
ISBN: 978-1-032-21305-7 (pbk)
ISBN: 978-1-003-26802-4 (ebk)

DOI: 10.1201/ 9781003268024

Typeset in Minion
by codeMantra

Contents

Preface

M ATHEMATICS IS KNOWN TO be difficult and not interesting to many people. However, some people including me think that mathematics is an interesting and challenging subject. Why is this so? We want to know the reason why the viewpoints of mathematics are different among people. I decided to write this book with the hope that people can change their negative mind on mathematics into the positive mind.

People who think that mathematics is exciting have a good feeling about mathematics since their childhood. Mathematics is made up of numbers. Kids already know or are familiar with numbers before they go to school. Kids eat an apple and ask for another one. They share some candies with their friends and count the remaining candies. They experience numbers in real life and learn formal numbers at school. Hence, they feel comfortable with simple numbers.

However, when they learn even and odd numbers, addition or multiplication of two-digit numbers, the patterns of various figures, and more advanced mathematical concepts as time goes by, they are gradually losing their interest in mathematics.

Then, is there a way to think that mathematics is creative and challenging by enjoying mathematical concepts? If we remember our old days with no internet, there were offline games. The most common game was played with marbles. If a player guesses the correct number of marbles hidden in a hand, the player wins the game. Sometimes, a player tries to hit marbles from a distance in order to send them further than others. In some sense, the former game is like an Algebra and the latter game is like a Geometry. By playing these kinds of games, I became familiar with mathematics and majored in mathematics in the end.

Thanks to this experience, I came up with games related to mathematics. I hope that adults as well as kids can eliminate their anxiety on mathematics by this approach and regard mathematics as a kind of game or play.

I have selected famous games for this book which also have some mathematical concepts. This book is based on a series of columns which appeared as "Jon-Lark Kim's board game festa" in the public magazine *Math Donga of Korea*. It consists of games which can be played alone or with other people. Some games can be played online. These games can be used as a means of teaching mathematical concepts at school or after school. Of course, they can be used for recreation among friends without any difficulty to learn mathematics hidden in the games.

In this book, there are games which help to learn basic mathematical concepts, and there are also games which contain very complicated mathematical concepts although the rules of the games are easy to follow. There is no problem enjoying the games even if you do not know the principles of mathematics. However, if you know the mathematical concept, you will enjoy the game with a lot of interest. There are mathematicians who try to solve the games mathematically. While there are also mathematicians who create mathematical games, I will try to introduce mathematicians who worked in various eras together with their brilliant ideas containing mathematical concepts. These mathematicians really enjoyed the games. With the games in this book, you can feel the taste of mathematics real mathematicians feel.

I was supported by many people. In particular, I thank my daughter Sylvie Jinna Kim for playing some of the games together when she was young and my wife Seung Kyung Yang for her encouragement. I am also thankful to Dr. Young Gun Roe and Hyeck Ki Min for the comments on the manuscript and Seung Jun Mun for drawing various figures.

The first fourteen chapters are basically the translation of the Korean book, and the remaining three chapters are added for this English version. I really appreciate the editorial team of the publisher, especially Mansi Kabra and Callum Fraser for their patience and advice.

If there are questions or comments on this book, please email me at ctryggoggo1@gmail.com.

Jon-Lark Kim from Seoul

CHAPTER 1

15 Puzzle

A Tile Matching Game That Is Difficult Even for Computers

A PUZZLE MADE BY A POSTMAN

The *15 puzzle*, also called *Boss puzzle*, is a game created in 1874 by Noyes Palmer Chapman, a postmaster in the town of Canastota, USA. In 1880, it became a sensation in the United States and Europe. About 130 years later, it is still loved by many people around the world. The *15 Puzzle* is also a game familiar to computer programmers. It is not known yet what

DOI: 10.1201/9781003268024-1

FIGURE 1   The *15 puzzle.*

algorithm is the best. Hence, many people are trying to find it. The mathematics of a 'permutation' is hidden in the game (Figure 1).

In the *15 puzzle*, square tiles numbered from 1 to 15 are placed side by side in a 4×4 square frame. A total of 16 tiles can be placed on the square frame and one spot is empty so that other tiles can move. First, place randomly the 15 numbered tiles and arrange the numbers on the tiles in an ascending order. This arrangement of numbers in order is called the 'standard array'.

In 1878, an American chess player and puzzle expert Sam Loyd (1841–1911) offered $1,000 to a person who would solve the puzzle where the 14th and 15th tiles were swapped (Figure 2). In other words, the puzzle is to go back to the standard array from an arrangement where the positions of tiles 14 and 15 from the standard array are switched. Due to such a prize, the *15 puzzle* became very popular in the United States. Loyd did a great job of making this puzzle popular with the public. In fact, if it is analyzed using a mathematical theory, it is shown to be impossible to convert this array into the standard array.

PERMUTATION

To figure out which arrays can be turned into the standard array or not, we must first represent them as a 'permutation'. A permutation is a function that represents an array of n elements, such as numbers or alphabets. Usually, n elements produce a total of $n!(=n\times(n-1)\times(n-2)\times\cdots\times2\times1)$

FIGURE 2   Illustration from Sam Loyd's book.

permutations. For example, consider the case of $n=3$, and the three elements are 1, 2, 3. If we consider all the arrangements of 1, 2, and 3, then 1–2–3, 1–3–2, 2–1–3, 2–3–1, 3–1–2, 3–2–1. There are a total of six permutations. If the blank space in the 15 puzzle is the 16th tile, there are 16 numbers in total, so there are $16! = 20,922,789,888,000$ permutations.

Now let us pick one of several permutations to show it in more detail. Comparing the tile arrangement of the *15 puzzle* in Figure 3 with the standard array, the table below can be represented. (This is called a one-to-one correspondence function.)

standard array	1	2	3	4	5	6	7	8	9	10	11	12	13	14	15	16
a given arrangment	1	2	3	4	5	7	10	8	9	6	16	11	13	14	12	15

In this table, except for the tiles 6, 7, 10, 11, 12, 15, and 16, all the other tiles are placed in the standard array. The 7th tile is placed on the 6th tile, the 10th tile is on the 7th tile, and the 6th tile is placed on the 10th tile. More simply, we can indicate which number is placed in the standard array position with an arrow like this:

$1 \rightarrow 1, 2 \rightarrow 2, 3 \rightarrow 3, 4 \rightarrow 4, 5 \rightarrow 5, 6 \rightarrow 7, 7 \rightarrow 10, 8 \rightarrow 8, 9 \rightarrow 9, 10 \rightarrow 6, 11 \rightarrow 16,$
$12 \rightarrow 11, 13 \rightarrow 13, 14 \rightarrow 14, 15 \rightarrow 12, 16 \rightarrow 15$

As shown above, if we use an arrow to explain which number is placed in the positions of tiles 6, 7, and 10 in the standard array, we can write $6 \rightarrow 7$,

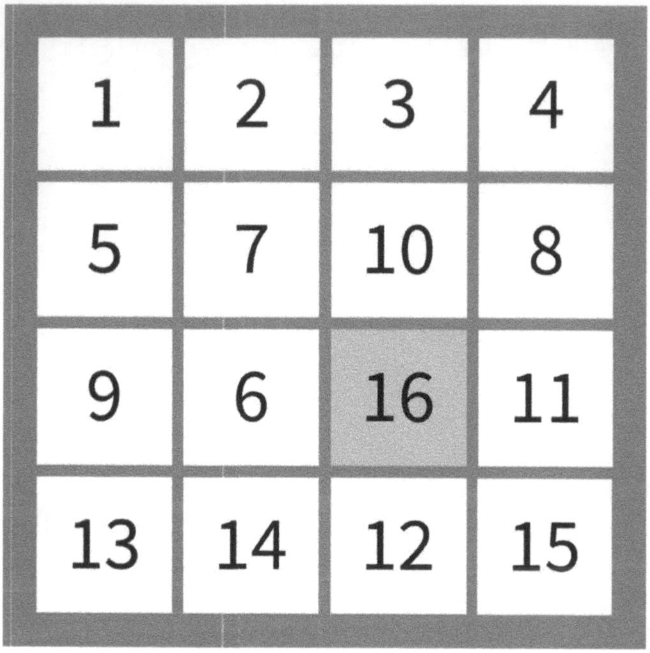

FIGURE 3 An arrangement of the 15 puzzle

7 → 10, 10 → 6. Now, write (6 7 10) except for the overlapping numbers and arrows. By reading from the left to the right, we think 7 in the place of the 6th tile, 10 in the place of the 7th tile, and 6 again in the place of the 10th tile. Similarly, the rest of the tiles can be represented as 11 → 16, 16 → 15, 15 → 12, 12 → 11, so we can write (11 16 15 12). Then these simple notations (6 7 10), (11 16 15 12) can represent an array.

EVEN PERMUTATION AND ODD PERMUTATION

When an array is expressed as a permutation, an odd number of numbers in parentheses is called an *even* permutation, and an even number is called an *odd* permutation. If the number is odd, why is it called an even permutation and not an odd permutation? The reason is not because of the number of numbers in the parentheses but the number of transpositions in the permutation. A transposition is an odd permutation of two numbers like (6 7), which can be expressed as 6 → 7, 7 → 6, so it is a symbol to swap the positions of the two numbers in parentheses. Any even or odd permutation can be expressed as a product of transpositions.

For example, (6 7 10) is expressed as (6 10)(6 7) by using two transpositions. This symbol means that in the standard array apply the transposition on the right first to swap the places 6 and 7, and then swap the places 6 and 10 again. That is, in the first (6 7), it is calculated as 6 → 7. There is no 7 in (6 10), so finally 6 → 7. How about 7? From (6 7) we have 7 → 6. However, from (6 10), we have 6 → 10, so if we connect these two, we end up with 7 → 10. Finally, what is the output value of 10? (6 7) has no 10, so it passes and instead (6 10) has 10 and 10 → 6, so 10 → 6 in the end. Similarly, (11 16 15 12) can be written as (11 12)(11 15)(11 16) using three transpositions. After all, the permutation represented by (6 7 10)(11 16 15 12) has even and odd transpositions, respectively, so the total is an even permutation+odd permutation=odd permutation.

Now we introduce our main theorem which states that under what condition a given array in the 15 puzzle can be moved to a standard array.

The only necessary and sufficient condition for a given array with the lower right corner spot empty in the 15 puzzle to be the standard array is that if the permutation corresponding to the given array is an even permutation. Only 16!/2 arrays can be moved to the standard array since the number of even permutations is equal to that of odd permutations.

This theorem was proved by American mathematicians William Johnson and William Story. According to this theorem, if the permutation representing an array is an odd permutation, it cannot be replaced with the standard array. Since Sam Loyd's array in which tiles 14 and 15 are switched can be expressed as odd permutation (14 15), it was an unsolvable problem actually. Let us see if the standard array is possible even in the case shown below. We represent this as a function:

standard array	1	2	3	4	5	6	7	8	9	10	11	12	13	14	15	16
a given arrangment	1	3	5	7	9	11	13	15	2	4	6	8	10	12	14	16

The corresponding permutation has the following form: (2 3 5 9)(4 7 13 10)(6 11)(8 15 14 12)=odd permutation+odd permutation+odd permutation+odd permutation=even permutation.

So this array can be the standard array. Of course, doing this in practice takes a lot of time. In general, there is a necessary and sufficient condition to become the standard array even for an $n_1 \times n_2$ ($n_1, n_2 \geq 2$) board.

Assume that the lower right corner spot is empty on the board $n_1 \times n_2$ ($n_1, n_2 \geq 2$). Then a necessary and sufficient condition for a given array in the $(n_1 n_2 - 1)$ puzzle to be the standard array is that the permutation corresponding to the given array is an even permutation.

MOVING TO THE STANDARD ARRAY WITH SMALLEST NUMBER

Using the theorem demonstrated by Johnson and Story, we can see which arrays cannot be standardized in the *15 puzzle* as well as in puzzles of various sizes. However, even if it is an array that can be moved to the standard array, it is not yet known how to move specifically which tiles and how to create a standard array by moving only the smallest number of tiles.

In 1986, American computer scientists Daniel Ratner and Manfred Watmuth found a way to create the standard array by moving only the fewest number of tiles from an $(n^2 - 1)$ puzzle consisting of n tiles in each column and row. They proved that the problem belongs to NP-hard. In other words, if you do an $(n^2 - 1)$ puzzle on an n×n board, it is very difficult to find the fastest way. (NP-hard is a set of problems with similar complexity of an algorithm for solving problems. There are P, NP, NP-hard, and PSPACE. NP-hard is the second most difficult set of problems after PSPACE.)

But we can tell at least how many times you need it this way: Let us look at the example (Figure 4). Using 'Manhattan Street' and 'Hamming Street' you can guess at least how many times a tile needs to be moved.

The Manhattan distance is the sum of the distances each number in this array must move horizontally or vertically to get to the standard array. That is, 1 is in the position of the standard array, so there is no need to move it; 3 is in the second position, so move it to the right once; 5 is in the

1	3	5	7
9	11	13	15
2	4	6	8
10	12	14	16

FIGURE 4   A special arrangement of the 15 puzzle

third position, so it needs to move down once and twice to the left for a total of 3 moves. This can be represented in a table as follows.

	1	2	3	4	5	6	7	8	9	10	11	12	13	14	15
Distance	0	3	1	4	3	2	2	1	1	2	2	3	4	1	3

So in this case you need to make at least 32 moves. Finding the Manhattan distance takes some time, so Hamming distance is efficient. You can find the distance and quickly get a rough idea of how many moves you need to move. Hamming distance gives a distance value of 0 if the given number of arrays is in the standard array, and a distance value of 1 otherwise. This is shown in a table as follows:

	1	2	3	4	5	6	7	8	9	10	11	12	13	14	15
Distance	0	1	1	1	1	1	1	1	1	1	1	1	1	1	1

So, you have to move at least 14 times to get to the standard array. Of course, the Hamming distance is always less than or equal to the Manhattan distance. In 1999, computer scientists Adrian Bürger et al. were able to roughly calculate the minimum and maximum values, proving that any array can be constructed in 80 times as long as it is an even permutation.

PROBLEMS

1. (*) Show directly (without any theory) that the left one cannot be moved to the right one.

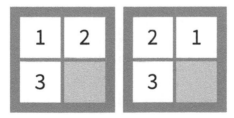

2. (*) Show that the array below has a permutation (1 9 8 7 6 5 4 3 2) and find a way to make the standard array. Also, how many moves are needed by the Manhattan distance?

	1	2
3	4	5
6	7	8

3. (**) Determine an even or odd permutation for the following array.

	1	2	3
4	5	6	7
8	9	10	11
12	13	14	15

4. (**) The following array has the property that the sum of numbers horizontally, vertically, and diagonally, respectively, is exactly 30. Determine whether this array can be moved to the standard array.

15	1	2	12
4	10	9	7
8	6	5	11
3	13	14	

5. Enjoy online *15 puzzle* at the following.
   http://lorecioni.github.io/fifteen-puzzle-game/
   * denotes the degree of difficulty.

CHAPTER 2

Peg Solitaire

Jumping Game Where We Meet Algebra

PEG SOLITAIRE WHICH CAN BE DONE ALONE

Peg solitaire is a board game played over 300 years. Peg means a 'stack', and solitaire means 'patience'. Usually one calls a one-player game a solitaire. But *Peg solitaire* can be played between two players. *Peg solitaire* was played by high-class French people in the 17th century. For example, Madam

DOI: 10.1201/9781003268024-2

FIGURE 5 Anne de Rohan-Chabot Princesse de Soubise (1663–1709) with *Peg Solitaire*, portrait painting, 1697.

Anne de Rohan-Chabot was playing it as seen in the 1687 French journal (Figure 5).

The rule of the game is simple. Place several pegs on a cross board. Pick one peg and jump it over a *neighbor* peg horizontally or vertically. The jumped peg is removed from the board. If there remains only one peg, the game is over.

Since *peg solitaire* has a long history, the shapes of the board vary depending on countries. There are three well-known shapes of the board. Two of them use the cross shape. The English board has 33 holes while the French or European board has 37 holes. The third one has a shape of a regular triangle with 15 holes.

Hence, the English *peg solitaire* has 33 holes and 32 pegs. We fill in the 32 holes except for the center hole with 32 pegs. Using the game rule, a peg can jump any neighboring peg, that is then removed from the board.

FIGURE 6   English solitaire board (above) and European solitaire board (below).

When this is played by one person, the person wins if there is only one peg left. When this is played by two persons, a person loses the game if the person cannot move. This book discusses the case when only one person plays the game (Figure 6).

WHERE DO WE PUT A LAST PEG?

Let us take a look at *peg solitaire* mathematically. For our convenience, the position of a hole can be described by the Cartesian coordinate system. Each location of a *peg solitaire* with x, y or z is given in Figure 7.

We explain how we place x, y, and z.

Any three consecutive holes correspond to the three different letters, and we place same letters along any anti-diagonal direction, or the right upper direction. The central hole is assigned to $(0,0)$. Then x, y, z, and the number 0 satisfy the following addition rule. This is called the *Klein four group* in Algebra. The equation $x + y = z$ means that x jumps over y to arrive at the position z.

+	0	X	Y	Z
0	0	X	Y	Z
X	X	0	Z	Y
Y	Y	Z	0	X
Z	Z	Y	X	0

When we play the English *peg solitaire*, the central hole (*y* value) is assumed to be empty in the beginning. To win the game, only one peg should remain at the end. What is a possible position of the last peg? Interestingly, the position occurs at one of the eleven holes where *y* is

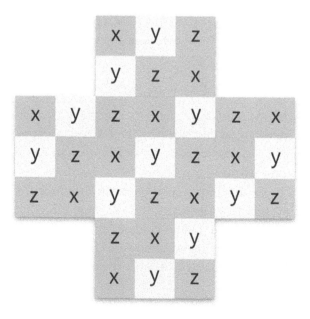

FIGURE 7   Value of each hole.

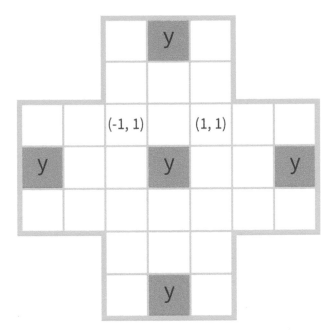

FIGURE 8 The possible positions for the last peg.

denoted. In particular, a peg can be located at the central position or one of four locations on each edge (Figure 8). The reason is as follows. If we reflect the y value at (1,1) about the vertical line of symmetry, we get $(-1,1)$ that has the z value. This violates the condition that the final value of the game should be y. Therefore, the (1,1) position is not a possible location for the final peg. Similarly, the four positions such as $(-1, -1)$, $(-1, 2)$, $(2, -1)$, $(1, -2)$ are not possible locations for the final peg.

The English *peg solitaire* starts with an empty hole in the center. If we start with an empty hole in other places, how many different games are possible? Clearly, there are 32 more games since there are 32 other holes except the center hole. It is natural to consider various symmetries. In this case, we need to consider the vertical and horizontal lines of symmetry together with rotations. Considering the symmetries, we can start with empty holes at x, y, z above the vertical line passing through the center and three additional locations that are at the left of these three holes. These cases produce six more non-symmetric games. Therefore, considering the original English *peg solitaire* game, there are seven different *peg solitaire* games based on the English board.

From now on, let us find a solution of the English *peg solitaire*. From Figure 9, there are three pegs in the block denoted by 1, which can be

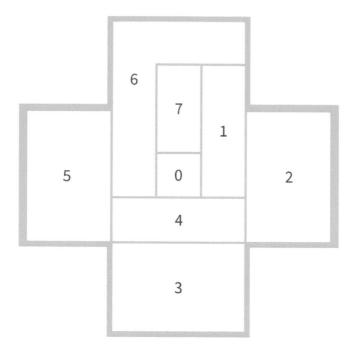

FIGURE 9   Finding solutions of English *peg solitaire* based on the orders of the block.

removed by using the central hole. There are six pegs in the block denoted by 2, which also can be removed successively. Following this pattern, there remain two pegs in the block denoted by 7, from which the last peg can be placed in the center (Figure 9).

TRIANGLE *PEG SOLITAIRE*

There are 15 holes and 14 pegs in triangle *peg solitaire*. In general, the top hole is empty. You can jump over an adjacent peg (which shares the edge) and remove the jumped peg. If there is only one peg left at the end, you win the game. Just like the English *peg solitaire*, we put x, y, z in the board. Assuming that there is only one peg left, where is the position of the last peg? The answer is one of the five positions denoted by x (Figures 10 and 11).

If the empty holes are one of the three green positions, then we have solutions in Figure 12.

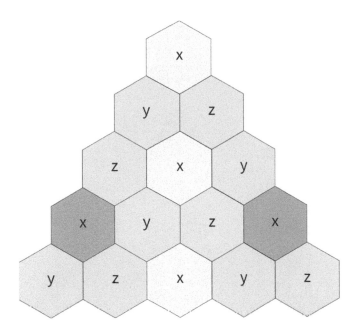

FIGURE 10   *x, y, z* labeling.

FIGURE 11   **Triangle *peg solitaire*.**

FIGURE 12　Empty position is at the top (A), the center (B), and the bottom (C).

PROBLEMS

1. (**) Try the English *peg solitaire* by yourself and find your own solution.

2. (**) Try the triangle *peg solitaire* by yourself and find your own solution.

3. (***) We are given the following peg. One of the orange positions is empty, and the other is filled with a peg. Solve the problem until the last peg is placed in one of the orange positions.

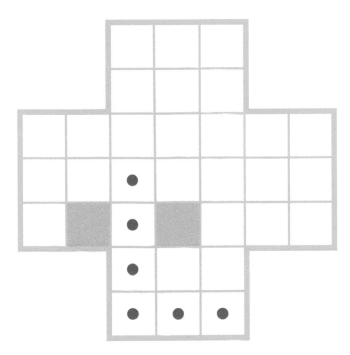

(Answer: Assume that the left orange position has a peg, denoted by number 0. We label the dot from the top to bottom as 1–6. Then jump the pegs as follows.

$0 \to 4 \to 1 \to 6 \to 3$. Then the peg on the right orange can jump the peg on position 2 to return the left orange position.

CHAPTER 3

Chomp Game

Avoid a Poisoned Chocolate

DO NOT EAT A POISONED CHOCOLATE BAR

Chomp is made by an American mathematician and economist David Gale (1921–2008). Gale studied game theory and Ramsey theory and also invented *Bridge-It* besides *Chomp*. *Chomp* means chewing food vigorously and noisily. As the name implies, two players eat chocolate bars alternately. The person who eats a poisoned chocolate bar will be the loser. You can also enjoy the game by using blocks or go stones instead of chocolate bars if the chocolate bars are unavailable.

DOI: 10.1201/9781003268024-3

The rule is simple. Prepare rectangle chocolate bars and assume that the bottom left piece has the poison. Each player chooses and eats a chocolate piece and additionally eats all the pieces that are above and on the right of the chosen piece. Players play the game alternately and if a player eats the poisoned chocolate bar, that person loses the game.

Let us play *Chomp* where the chocolate has four pieces vertically and five pieces horizontally as in Figure 13. Player A eats the right six pieces. Player B eats the right two pieces. Continuing this, B will leave the poisoned piece at the 6th step, and therefore A will lose the game. Play *Chomp* that has the same size with your friends.

If the size of the chocolate bar is $m \times n$ then it is called an $m \times n$ *chomp*. The position of the poisoned chocolate bar is assumed to be at the bottom left. We can represent it as the Cartesian coordinate (1,1) and represent the positions of the other bars in a similar way. For example, the bar with the third position horizontally and the second position vertically is denoted by (3,2). Therefore, when the (p,q) bar is chosen, a player should eat all of the pieces that are located at (r,s), where $r \geq p$, $s \geq q$.

Since the two players play alternately, it appears that there is no winning solution. However, Gale proved in 1974 that the first player can win the

FIGURE 13   **An example of Chomp.**

game for any size of the chocolate bar. To explain the proof, suppose that I choose (p, q) piece and the other player chooses (i, j) piece.

We can think of two cases. In the first case, if *I* win the game after *I* choose (p, q), then this position is the winning position. In the second case, *I* lose the game after *I* choose (p, q). The reason is that the other player has chosen the position (i, j). Hence if *I* chose (i, j) in the beginning, then *I* would win the game. This proof is called the existence proof. But except for some *m, n*, it is not known how to choose positions concretely.

Now let us find a simple winning strategy. Consider $1 \times n$ *chomp*. If I choose (1,2) and eat all the bars to the right, then the other player should eat (1,1) and so I win the game. What happens if we play $m \times m$ *chomp*? The answer is to choose (2,2) and eat all the bars above it and on the right. If the other player chooses $(1, j)$ or $(j, 1)$ then *I* choose $(j, 1)$ or $(1, j)$ symmetrically. Then in the end the other player has no choice but (1,1).

Now let us consider 3×2 *chomp*. The position *I* should eat is (2,3) (see Figure 14).

How about 3×4 *chomp*? (see Figure 15). First *I* choose (3,2). Then think about the other positions.

We might think about the case when *m* or *n* is infinite. For convenience, assume that *n* is infinite(∞). $1 \times \infty$ represents an infinite chocolate bar. To win the game, it is enough to choose (1,2). In $2 \times \infty$ *chomp*, the other player can always win the game (try this!). If *m* is greater than or equal to 3, then I can always win $m \times \infty$ *chomp*. The reason is that if I choose (1,3), the other player should choose a position on $2 \times \infty$ *chomp*.

FIGURE 14 3×2 *Chomp*.

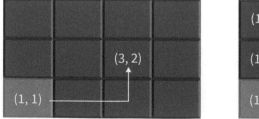

FIGURE 15 3×4 *Chomp*.

PARTIALLY ORDERED SET

Chomp game can be understood as a partially ordered set. Partially ordered set is a set where some elements of the set have an order.

{1, 2, 3, 4, 5}	{1, 2, 3, a, b}

Any two elements in the first set {1, 2, 3, 4, 5} can be compared. However, *a* and 1 in the second set {1, 2, 3, a, b} cannot be compared since *a* is not a number.

Let us transform 2×3 *chomp* into a partially ordered set in a simple way. First, we represent each chocolate piece as a coordinate and make a set {(1,1), (1,2), (2,1), (2,2), (3,1), (3,2)}. In this set, (*a,b*) is called *greater than* (*c,d*) if (i) *a* is greater than *c* and *b* is greater than or equal to *d* or (ii) *a* is equal to *c* and *b* is greater than *d*. For example, (2,2) is greater than (2,1) but it is not compared with (3,1). Thus the set is a partially ordered set. Now instead of choosing a chocolate piece, choose a coordinate and instead of eating a chocolate bar, erase all the coordinates greater than or equal to the chosen coordinate. Then, the person who erases (1,1) will lose the game.

$m \times n$ *chomp* can also be explained as a 'factor game'. Let's consider the number $N = 2^m \times 3^n$. Write down all the (positive) divisors of N of the form $2^a \times 3^b$, where *a* and *b* are at least 1. For example, 2×3, 2×3^2, $2^2 \times 3$, $2^2 \times 3^2$, ... are possible divisors. Instead of choosing a chocolate piece, we choose one of the divisors and erase all the multiples of this divisor so that the person who erases 2×3 will lose the game.

3-DIMENSIONAL *CHOMP*

So far I have explained the 2-dimensional *Chomp*. A 3-dimensional *Chomp* is also possible. Make a chocolate block instead of a chocolate bar and consider a cuboid piece instead of a rectangle piece. The poisoned piece of a cuboid is one of the corner positions and let (1,1,1) be the position. Then the cuboid is defined as the set of (*i, j, k*) where $1 \leq i \leq m, 1 \leq j \leq n, 1 \leq k \leq l$, and we can define an order in the set. If $i' \leq i, j' \leq j, k' \leq k$, then $(i', j', k') \leq (i, j, k)$. We do not define the order for other cases, and hence this is a partially ordered set. Figure 16 is a $4 \times 4 \times 4$ *chomp*. The person who chooses the red block loses the game. If a yellow block is chosen, the green blocks are erased.

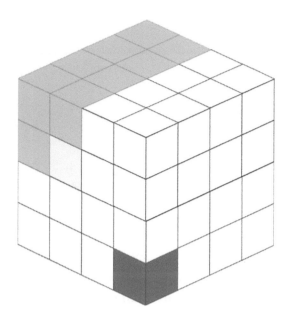

FIGURE 16 3-dimensional Chomp.

NIM is similar to *Chomp*. Instead of the chocolate pieces, one or more beads will be moved simultaneously from one of the three piles of the beads hanging on a string. Two players alternately move the bead(s). If one cannot remove a bead, the person loses the game.

PROBLEMS

1. (*) In 2×3, 2×4, 2×5 *chomp*, if you want to win the game, which piece should you choose? If you find the piece, try a solution for a general 2×*m chomp* game, where *m* ≥ 2.

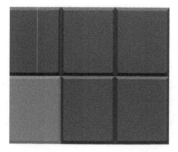

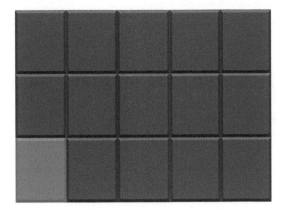

2. (**) In a $2 \times \infty$ *chomp* (horizontally infinite), what is a winning strategy that the second player wins the game?

3. (**) In a $3 \times m$ *chomp* ($5 \leq m \leq 12$), find a position where the first player wins the game.

4. (***) In a $3 \times 3 \times 3$ *chomp*, find a position where the first player wins the game.

5. (****) In a $3 \times 3 \times \infty$ *chomp* (height is infinite) and the $\infty \times \infty \times \infty$ *chomp*, find a position where the first or second player wins the game.

CHAPTER 4

Super *Tic-Tac-Toe*

Stone Game over a Donut

TIC-TAC-TOE GAME

In Korea, there is a game called the 5-stone game. In the Go board, the first player picks a stone of one color and the second player picks the other color. They put their stones alternately and one wins the game if the person makes five stones of the person's color connected horizontally, vertically, or diagonally.

DOI: 10.1201/9781003268024-4

In this sense, the *tic-tac-toe* game can be thought of the 3-stone game. Instead of putting stones, *tic-tac-toe* uses X or O alternately on the 3×3 board. The person who makes three X's or O's wins the game. *Tic-tac-toe* has been played from the ancient Egypt. Because the rule is so simple, the game ends tie often when the two experienced players play. However, there is a more interesting *tic-tac-toe* like game.

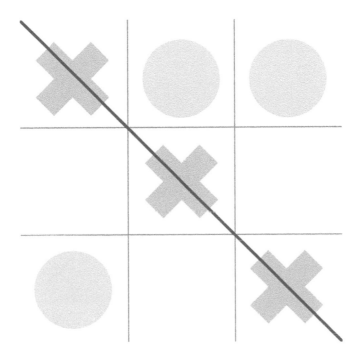

A NEW CONNECTION, AFFINE PLANE *TIC-TAC-TOE* GAME

In 2006, an American mathematician Stephen Dougherty suggested an interesting *tic-tac-toe*. He adds more three positions for a win as follows.

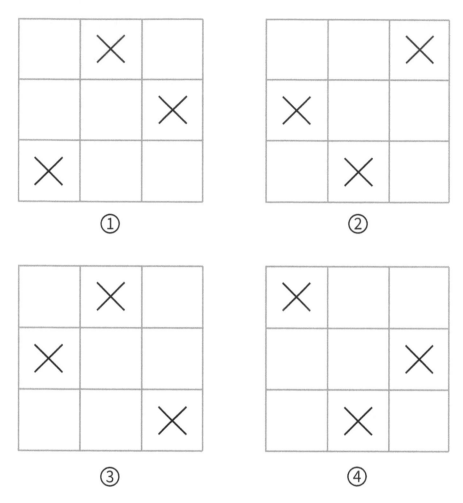

What is the property of this *tic-tac-toe*? Let us look at ①. The first two rows mark X diagonally to the right and if we extend it then it will be like Figure 17. The third X is then off the 3×3 board. However, if we assume that the 3×3 board is connected continuously, then the position corresponds to the bottom lower left of the 3×3 board. Therefore, the three Xs in ① is another set of winning positions. Similarly, even though the three Xs in the remaining cases like ②, ③, and ④ look apart, they are considered to be connected as in Figure 17. We call this *tic-tac-toe* 'affine plane *tic-tac-toe*' or 'torus *tic-tac-toe*'. Torus is like a tube for swimming or a donut with one whole. The reason why it is called a torus *tic-tac-toe* is that if we roll the top and the bottom of the *tic-tac-toe* board to make a straw and then connect the end of the straw, then we obtain a torus. Therefore, the new three Xs in the affine plane *tic-tac-toe* are in fact connected on the torus (Figure 18).

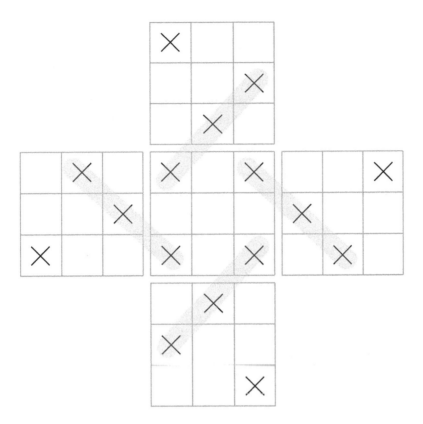

FIGURE 17   Diagonally connected *tic-tac-toe*.

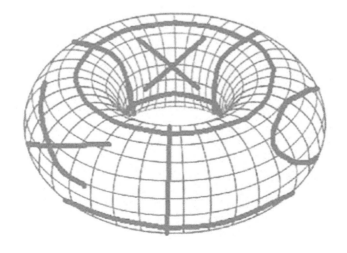

FIGURE 18   Torus *tic-tac-toe*, from codegolf.stackexchange.com.

The 3×3 *tic-tac-toe* is usually a tie game even if we fill the board. However, this is not true for the affine plane *tic-tac-toe*. Why is that so? Suppose that we have a tie game after we fill in the affine plane *tic-tac-toe*. Assume that the first player marks X. Then, the board has five Xs and four Os. Now, we can consider four cases where four, three, two Xs and one X are on edges of the board and prove that a tie game cannot occur in all of the four cases.

1. Suppose that Xs appear on the four sides as in Figure 19①. The spots where Xs appear are filled with a color. To have five Xs in total, the remaining X should be added to some empty spot. If X is marked at the center, we have two 3 Xs horizontally and vertically. If X is marked at any one of the four corners, they correspond to one of the cases in Figure 17. Therefore, X will win the game.

2. Suppose that Xs appear at three spots on the four sides as in Figure 19②. We need to add two more Xs in the board. If we place X in any of the positions marked as N, then X will win the game. Therefore, we have to consider two empty spots on the top row. If we put two remaining Xs in those two empty spots, we have three Xs in the top row. Therefore, X will win the game.

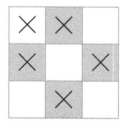
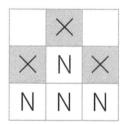

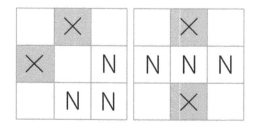
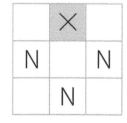

FIGURE 19 Affine plane *tic-tac-toe* winning case. Labeled as ①, ②, ③, ④ from the top.

3. Suppose that Xs appear on the four sides as in Figure 19③. Then there are two cases. As in 2, N is the position where X needs to be avoided. If three Xs are placed in any of the four empty spots, a three Xs will be formed and therefore X will win the game.

4. Figure 19④ has one X marked and four forbidden positions marked as N. In these cases, four Xs should be added. If one X is marked at the center, then the remaining three Xs are marked at three of the four corners, which results in a diagonally winning three positions. If one X is not marked at the center, then the remaining four Xs are marked at the four corners. This top row is the winning three positions. Therefore, X will win the game

Therefore, the affine plane *tic-tac-toe* game ends with a win. Even though the first player has some advantage, the game can end before all the Xs and Os are marked so the second play needs some strategy for a win.

AN ORDER FOUR AFFINE PLANE *TIC-TAC-TOE* GAME

The previous *tic-tac-toe* game is an order 3 affine plane *tic-tac-toe*. An order 4 affine plane *tic-tac-toe* is a *tic-tac-toe* game where we make a set of four connected stones on the affine plane of order 4 as in Figure 20. Each color has four sets of four connected stones. There are five colors. Therefore, there are twenty winning sets of four connected stones.

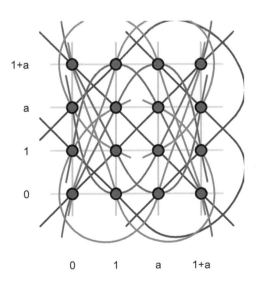

FIGURE 20   The order four affine plane *tic-tac-toe*.

SUPER *TIC-TAC-TOE*

Super tic-tac-toe or *Ultimate tic-tac-toe* is a game on a big 3×3 board where each cell has another *tic-tac-toe* 3×3 board. If X or O wins in a 3×3 cell, then mark X or O accordingly. The person who makes three Xs or Ox first overall on the big 3×3 board wins the game.

The rule is as follows. Choose one of the nine 3×3 *tic-tac-toes* in the 9×9 board (=the big 3×3 board) and then put X or O there (see Figure 21). For example, if the first player chooses the middle *tic-tac-toe* cell and put X at the right corner, then the second player moves to the upper right *tic-tac-toe* cell and puts O at any empty position. Then the first player moves to the corresponding *tic-tac-toe* cell and puts X there and so on. If a *tic-tac-toe* cell ends with a win or a tie, then skip that *tic-tac-toe* cell and move to another *tic-tac-toe* cell and put X or O.

It is not known whether there is a winning strategy for *super tic-tac-toe*. In 2013, Eytan Lifshitz and David Tsurel (AI Approaches to *Ultimate Tic-Tac-Toe*, preprint), PhD students at Computer Science Department, Hebrew University of Jerusalem, Israel used the computer simulation and analyzed the statistics to conclude that the winning probability that the first player wins the game is 56%. Recently in 2020, Bertholon et al. (At most 43 moves, at least 29: optimal strategies and bounds for *ultimate tic-tac-toe*, preprint) showed that there is a winning strategy for the first player.

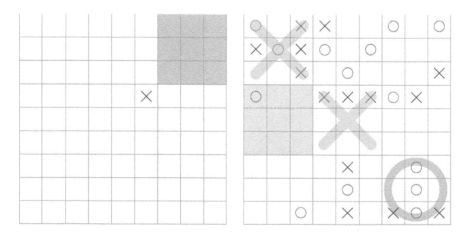

FIGURE 21 *Super tic-tac-toe.*

PROBLEMS

1. (*) Play the order 3 affine *tic-tac-toe* with your friends.

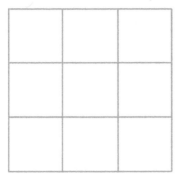

2. (*) Play *Super tic-tac-toe* with your friends.

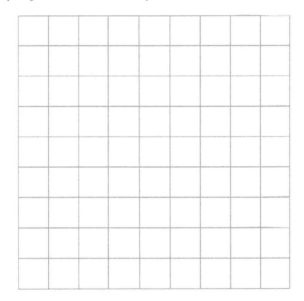

3. (**) Below is an order 4 affine plane *tic-tac-toe*. Show that it is a tie game.

X	O	X	X
O	X	O	O
X	O	X	X
O	X	O	O

CHAPTER 5

Mastermind

A Cryptographic Game

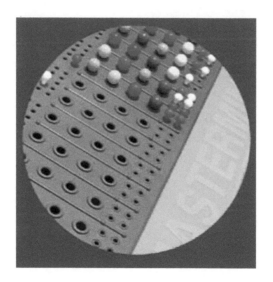

GUESS THE PASSWORD

Mastermind is a board game guessing the opponent's password or secret code. Mordecai Meirowitz, an Israel postmaster and telecommunications expert, invented the *Mastermind* in 1970. After Meirowitz invented the game, he contacted many game companies but they refused to commodify his invention. Luckily, a plastic company, Invicta Plastics, showed an interest and purchased all the rights to the game to be available on the market.

DOI: 10.1201/9781003268024-5

FIGURE 22   *Mastermind*, CC BY-SA 2.0.

A *Mastermind*'s board consists of several rows, each of which has four small holes and four large holes. The number of rows depends on the difficulty of the game (see Figure 22), usually being 12. On top of that, there are pegs of six colors, and there are small pins of red and white colors. One can put up to four pins in each row.

Mastermind is a two-player game. One person is called a codemaker and the other is called a codebreaker. Codemaker chooses four colored pegs out of the six colored pegs and hides them at the bottom of *Mastermind*'s board. We call them a password or secret code or code simply. The color of the four chosen pegs is allowed to be the same. Now the codebreaker should guess the color of the four pegs and place them in the first row. Then the codemaker puts a red pin (R) on the right if the color and its position are correct. If the color is correct but the position is not correct, then put a white pin (W). Otherwise, do not put any pins there.

Repeat this process up to 12 rows. If the codebreaker guesses the password correctly within 12 rows, the codebreaker wins the game. Otherwise, the codemaker wins the game. If one game ends, change the role. After fixing the total number of the plays, players with more wins will win the game.

ANALYZE *MASTERMIND* MATHEMATICALLY

It was Donald Knuth, an American computer scientist, mathematician, professor emeritus at Stanford University who analyzed *Mastermind* mathematically. He earned a PhD in Mathematics from Caltech in 1963 under his advisor Marshall Hall, a famous group theorist. At the age of 24, he published a book titled *The Art of Computer Programming*, which is regarded as one of the monumental books in the computer science area.

Knuth published a paper (The computer as master mind, *J. Recreational Mathematics*, Vol. 9, No. 1, p. 1–6, 1976–1977) on *Mastermind* in the *Journal of Recreational Mathematics* in 1976. He represented those six colors as 1, 2, 3, 4, 5, 6 and four pegs as four numbers with repetition allowed. Therefore, there are $6^4 = 1,296$ possible passwords. He made an algorithm to suggest numbers in each step and proved that the password can be found within at most five guesses.

Because Knuth's proof is complicated, let us explain it with a simple example. Suppose that you are a codebreaker and assume that a codemaker picked a password consisting of four numbers from 1 to 6. We try 1122 first. If there are four white pins(write '4W'), then the only possible answer is 2211. If there are three white pins (3W), then 221□, 22□1, 2□11, □211 (here □ can be one of 3, 4, 5, 6 except 1 and 2). Thus there are 16 ways to consider. To test these 16 cases, we ask 1213 which looks unrelated (see Figure 23). Then we will get at least 1R. Among them, if 1R2W is the

□ =	3	4	5	6
221□	3R	2R	2R	2R
22□1	1R2W	1R1W	1R1W	1R1W
2□11	1R3W	1R2W	1R2W	1R2W
□211	2R2W	2R1W	2R1W	2R1W

FIGURE 23   The case of 1213.

answer, 2231, 2411, 2511, and 2611 are passwords. Then let us ask 1415. Then there are four cases as follows:

2231 1W

2411 2R1W

2511 1R2W

2611 1R1W

By looking at this output, we can guess an original password. Even if 1R2W is not an answer, we can find an answer in a similar way. If 3R appears out of the 16 cases, then as it appears once as in Figure 23, the corresponding password is 2213. There are three cases that 2R appears out of 16 cases, the password is one of 2214, 2215, and 2216. We may simply take one of them, but then we may ask up to three questions. To reduce the number of questions, we ask 1415 as in the case of 1R2W. Then 2214 corresponds to 1R1W, 2215 to 2R, and 2216 to 1R, from which we can find the password. In the case of 1R1W, we also ask 1415 so that we can find the password. Altogether, we can find the password with up to four questions.

In the above, we have proved that if there is 3W, it suffices to ask four questions. What happens if there is 2W? In this case, the possible passwords are as follows and we have $4^2 \times 6 = 96$ cases.

22☐☐, ☐☐11, 2☐1☐, ☐2☐1, 2☐☐1, ☐21☐

(here, ☐ denotes one of 3,4,5,6.)

If there is one white pin (1W), we have $3^4 \times 4 = 324$ cases.

☐☐1☐, ☐☐☐1, 2☐☐☐, ☐2☐☐

(here, ☐ denotes one of 3,4,5,6.)

If there is no white pin (0W), then each position should be one of 3,4,5,6 which results in $4^4 = 256$ cases.

GUESS THE PASSWORD IN 4.34 QUESTIONS

Many researchers besides Knuth tried to find an algorithm to find the password of *Mastermind*. In 1993, Kenji Koyama and Tony W. Lai (An optimal Mastermind Strategy, Journal of Recreational Mathematics, 25(4):251–256, 1993) showed that there is a method to find the password with the average of 4.34 questions. According to this paper, at most six questions are enough to break the secret code. Jeff Stuckman and Guo-Qiang Zhang (Mastermind is NP-complete, INFOCOMP Journal of Computer Science 5, 25–28 (2006)) proved that the problem to decide whether given a random set of colored pegs and its score, there is at least one password satisfying these conditions is a NP-complete problem.

There are several games similar to *Mastermind*. Bulls and Cows is a pencil-and-paper game. This was a well-known game before the appearance of *Mastermind*. Bulls and Cows is a two-player game. One player plays the role of 'Bull' and the other plays the role of 'Cow'.

Bulls and Cows breaks the password consisting of four digits from 0 to 9. We do not allow the repetition of the digits, which makes the difference from *Mastermind*. Instead of putting pins, if the codebreaker guesses the correct number and its position, it is called 1Bull and if the codebreaker guesses the correct number and the codebreaker guesses incorrect position, then it is called 1Cow. For example, suppose that the password is 1357 and the codebreaker says 1235. Then the number 1 is correct and its position is also correct so we have 1Bull. The numbers 3 and 5 are correct but their positions are different from the password so we have 2Cows. Because Bulls and Cows was very popular, a student at MIT in the late 1960s made a game program called 'MOO'.

There is also word *Mastermind*. This game uses four alphabets instead of numbers as password. The rules are the same as those of Bulls and Cows. If FOUR is the password and the codebreaker guesses GOOD, then the codemaker answers with 1Bull and 0Cow.

Super Mastermind is another type of *Mastermind*. The rules of *Super Mastermind* are the same as those of *Mastermind*. There are eight colors of pegs and five holes for the password (Figure 24).

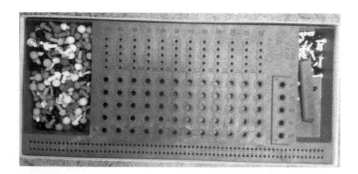

FIGURE 24 *Super mastermind.*

PROBLEMS

1. (*) Play *Mastermind* using numbers 1–6 with friends.

2. (*) Play *Mastermind* using numbers 0–9 with friends.

3. (**) Play word *Mastermind* using the first 10 alphabets.

4. (***) The following is part of *Mastermind* using numbers 1–6. What is the final password? Try this for 10 minutes without the hint: The password consists of one 2, two 3s, and one 6.

Questions	Scores
1122	1R
1344	1W
3526	1R, 2W
1462	1R, 1W

CHAPTER 6

Ramsey Theory and Sim Game

Don't Draw a Red Triangle

SIM, A GAME ON THE REGULAR 6-GON

Sim (or the game of SIM) is a pencil-and-paper game played by two players on the 6-gon. An American cryptographer Gustavus Simmons introduced this game in 1969 (The game of SIM, *J. Recreational Mathematics*, 2(2), 1969, pp. 66). Many people tried to find a winning strategy.

DOI: 10.1201/9781003268024-6

Sim is played by drawing lines on the regular 6-gon with all edges connecting any two vertices. Note that the regular 6-gon has six vertices. Hence there are 15 edges connecting any two vertices. For example, we represent the line connecting vertices A and B by line AB so that there are 5 lines starting from vertex A. Similarly, there are 4 lines starting from vertex B, 3 lines starting from vertex C, 2 lines starting from vertex D, and 1 line starting from E. Hence by adding numbers from 1 to 5 we get 15 distinct lines. We can also compute this number by counting all permutations on six objects (Figure 25).

The rule of *Sim* is very simple. Mark six vertices on a paper and each player draws one of the 15 lines alternately. Suppose that the first player (R) uses a red pen and the second player (B) uses a blue pen. If there is no colored pen, use a pencil with dots or straight lines. Players draw their colored lines alternately. If one player has a triangle with one color, that player loses the game. Thus, in order to win the game, each player should avoid a triangle with each player's color and make the other player draw a triangle with one color.

Let us give a simple example. In Figure 26, the left game has been progressed from a red line AC first and there are seven lines now. So far there is no triangle with one color. Now it is player B's turn. If player B draws line FD or line FB, there is a triangle with blue color and so player B loses

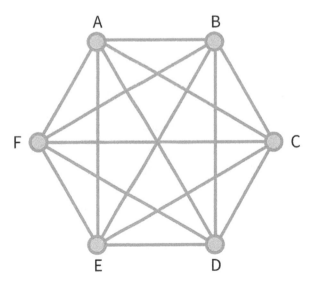

FIGURE 25 *Sim* game on the 6-gon.

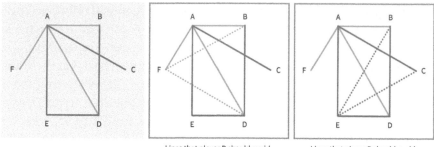

Lines that player B should avoid.   Lines that player R should avoid.

FIGURE 26   Lines that each player should avoid.

the game. Therefore, play B should find other lines. Similarly, what lines should player R avoid? These are line EC and line EB.

Now we continue to draw lines and ask whether it is possible that there is no triangle with one color at the end. I mean if *Sim* can end with a tie game. The answer is that the game cannot be a tie game. Frank Ramsey, a British philosopher and mathematician proved this.

RAMSEY'S THEOREM

Frank Ramsey was a genius and died prematurely. He was born in 1903 in the family of mathematicians and died at the age of 28. His favorite philosopher was Ludwig Wittgenstein who was a best analytical philosopher at that time. Ramsey translated Wittgenstein's <Tractatus Logico-Philosophicus> into English at the age of 20. He discussed Math and Philosophy regularly with Wittgenstein at Cambridge University.

Ramsey made an important contribution to Logic. However, what made him famous was Ramsey's theorem which says that there always exists a subgraph with a special property in a graph. Basic facts are as follows.

A graph with n vertices such that there is an edge between any two distinct vertices is called a complete graph, denoted by K_n. You may visualize K_n as a regular n-gon with all edges between any two vertices. For example, when $n = 3$, it becomes a triangle. When $n = 4$, it becomes a square with two more edges inside. When $n = 5$, it becomes a pentagon with a star inside. See Figure 27.

Theorem: *If all the edges of K_6 are colored by red or blue, then there exists a triangle K_3 with monochrome.*

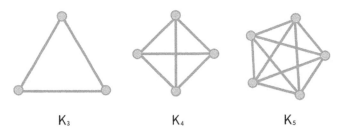

K_3 K_4 K_5

FIGURE 27   Complete graphs.

PROOF OF RAMSEY'S THEOREM USING FINGERS

Shall we prove Ramsey's theorem briefly using fingers? Represent the six vertices of K_6 by the five fingertips of a right hand and a middle point p of palm. See Figure 28. Connect p with those five fingertips. Then there are at least three lines with monochrome. Why is that so? There will be red or blue lines from p to the five fingertips. Since there are five lines,

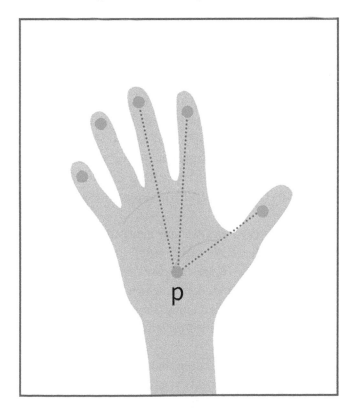

p

FIGURE 28   Proof of Ramsey theorem using fingers.

there should be at least three red lines or at least three blue lines. This is because if there are only two red lines and only two blue lines, then there are only four colored lines which is a contradiction.

Suppose that there are three red lines. Let us focus on the three fingertips. If there is a red line connecting two of these fingertips, then the red line and two previously colored red lines form a red triangle. If the three fingertips are connected with only blue lines, then we get a blue triangle. Thus in any case, we get a triangle with monochrome.

I will explain Ramsey's theorem using another example. In a party with at least six people, it is true that either three people know each other or three people do not know each other.

Consider a complete graph K_5 which has only five vertices. Can you find a triangle with monochrome? The answer is no. If you color the five inside lines in blue and five outside lines in red, you cannot find a triangle with monochrome. See Figure 29. Since *Sim* is based on Ramsey's theorem, we cannot play on a pentagon.

The minimum number n of vertices of K_n such that there exists a subgraph K_r in red color or a subgraph K_s in blue color is called a Ramsey number, denoted by $R(r,s)$. *Sim* game is based on Ramsey's theorem saying that $R(3,3)=6$. Although it is known that $R(r,s)$ exists, it is very difficult to find the exact value. Still after 40 years of Ramsey's theorem, it is only known that $R(5,5)$ is greater than or equal to 43 and is less than or equal to 48.

It is known that the second player (red) can have a winning strategy. The actual method is rather complicated and so there is no easy method yet to apply the strategy. One simple rule is that when one line is selected among the player's available lines try to keep unavailable lines to a minimum. Following this rule, the player with more lines available will win the game at the end.

If we use Ramsey's theorem, we can generalize *Sim* game to a graph with more than six vertices. We call it 'graph Ramsey game'. For example, using the fact that $R(4,4)=18$, we can play on K_{18} with the game rule that the player drawing K_4 with monochrome loses the game. If more Ramsey numbers are found, then we can enjoy *Sim* game on various graphs. The graph K_{18} in Figure 30 contains 153 edges so it will be complicated to play on paper.

In addition, we can play with three colors. It is known that $R(3,3,3)=17$. This means that there always exists a triangle in red, blue, or green color in the graph K_{17} and that there is a coloring of the edges with no triangle in monochrome in the graph K_{16}.

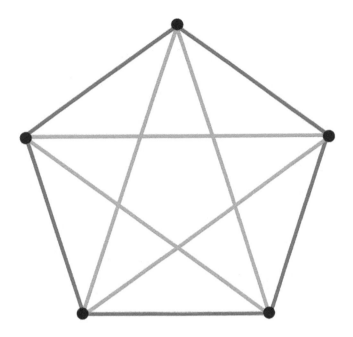

FIGURE 29   Is *Sim* game possible on a pentagon?

FIGURE 30   *Sim* game on the 18-gon.

PROBLEMS

1. (*) Let us play *Sim* game with friends. The game can be a tie.

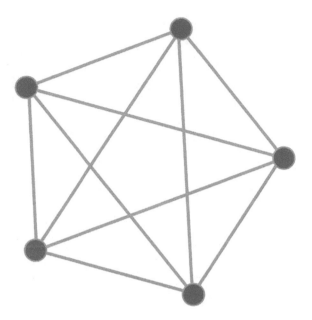

2. (**) It is known that $R(3,3,3)=17$. To prove this, we need to show that $R(3,3,3)\neq16$. In the following graph with the three colored edges, there does not exist a triangle with only one color. Check that there is no triangle with red color.

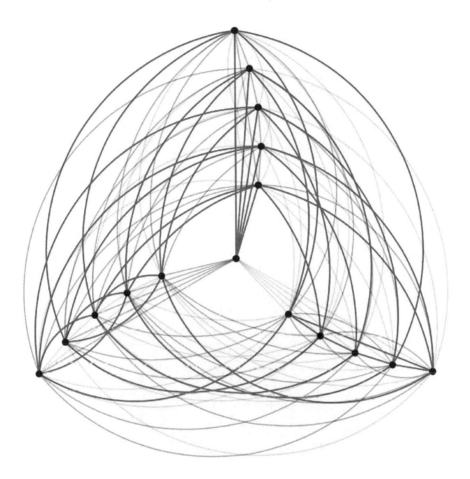

3. (***) There are several strategies to win *Sim* game. Nevertheless, an effective strategy in a real game is not known yet. Find your own strategy.

4. (****) Currently it is known that $43 \le R(5,5) \le 48$. Prove that either $R(5,5) > 43$ or $R(5,5) < 48$.

REFERENCE

1. Simmons, Gustavus J. "The game of SIM," *J. Recreational Mathematics*, 2(2), 1969, pp. 66.

CHAPTER 7

Nine Men's Morris

Three Soldiers Side by Side

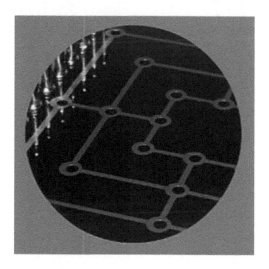

GAME ENJOYED IN THE ROMAN EMPIRE AND INDIA

Nine men's morris is an ancient board game played by two players since the Roman Empire. It has been played around the world including Egypt, England, and North America. Since it consist of pegs and a board, it has been one of the popular games. These days, one can also have fun with a web version of Nine men's morris. Because the word "morris" has the meaning of a "coin" and the shapes of the coin represent soldiers, it is called a *"nine men's morris"*.

 DOI: 10.1201/9781003268024-7

FIGURE 31   9 Men's Morris from the book <Libro de los juegos> in 1283.

To play *nine men's morris*, each player prepares the nine pegs with one color (usually black or white) or shape to distinguish with the other player on a board consisting of 24 points.

The game is simple. Two players put pegs on an empty place alternatively. When there are three pegs of the same color in a line, it is called a "mill". If one player makes a mill, the player can remove the other player's peg. A player loses the game if the player has only two pegs or cannot move a peg anymore. Shall we take a closer look at the game? (Figure 31).

FIRST STEP: PLACING PEGS ON A BOARD

Two players place their pegs alternately on empty spots out of the 24 points. If one player places three pegs in a horizontal line or in a vertical line, then the player can remove the other player's peg. Here, first try to remove the other player's peg not in a mill. If such a peg is not available, then the player can remove a peg from a mill.

SECOND STEP: PUSHING PEGS ON A BOARD

If all the pegs are placed in the first step, then the players need to push their pegs to make a mill. This time, a peg can be moved to an adjacent point. There is a tip. If a player can move a peg in a mill to an adjacent point and make a mill again, then the player can remove the other's peg. Continue this way. If a player has only three pegs left, move to the third step (Figure 32).

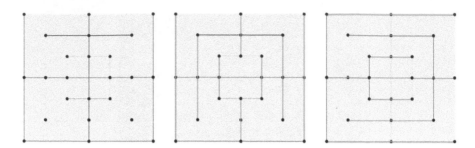

FIGURE 32   Mills on vertical lines and horizontal lines shown in blue color.

THIRD STEP: JUMPING A PEG ON A BOARD

If there are only three pegs left, it is like reaching a dead end. If the opponent player makes a mill and removes my mill, then it is not possible for me to make a mill. Therefore, if there are only two pegs from one of the two players, the game is over. If there are only three pegs left, a peg can jump a point and move to any empty point. This third step is not always necessary. If the players agree, then they can play until there are two pegs left based on the first two steps.

SEVERAL MORRIS GAMES

Although *nine men's morris* game is most common, other numbers of men are also possible. Three men's morris can be played on the board with nine points with three pegs, which is called '9 holes'. In this game, the player who makes a mill first (horizontally, vertically, or diagonally) wins the game. If two players have not made a mill, they can play based on one of the two options. The first option is to move a peg to an adjacent point as in step 2. The second option is to move a peg to any empty point.

Six men's morris has a board with 16 points and 6 pegs for each player. The rule is the same as *nine men's morris*. This game is more fun if the rules of the first step and the second step are used without the third step (Figure 33).

Twelve men's morris has more pegs and lines than *nine men's morris*. It is known as 'morabaraba' in South Africa. At first glance, the board looks the same but there are four more diagonal lines at the four corners so that one can make more mills. The rules of this game are the same as those of *nine men's morris*. The only different rule is that if both players place 12 pegs on the board without making a mill until then, the game is a draw.

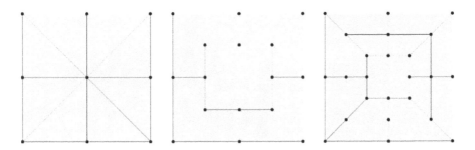

FIGURE 33   3 men, 6 men, and 12 men's morris in order.

WINNING STRATEGY FOR *NINE MEN'S MORRIS*

It is not known whether there is a winning strategy for *nine men's morris*. Instead, when the game has somewhat progressed, there is an analysis for the probability that the first player (Black pegs) wins based on the condition that where and how many pegs are left. In 1996, Ralph Gasser, a software engineer from Switzerland proved that the game ends in a draw in most cases by an exhaustive search. (See *Solving Nine Men's Morris, Games of No Chance. 29: 101–113. Retrieved 2015-06-01, 1996.*)

Gasser made a 3D bar graph for the probability that Black pegs win the game according to the number of White pegs and Black pegs left in second step. To understand the graph, we represent *a–b* when there are *a* Black pegs and *b* White pegs. For example, 7–4 means there are a 7 Black pegs and 4 White pegs. In this case, the game changes into 6–4 or 7–3 game. In other words, the current pattern of pegs is affected by the previous pattern of pegs (see Figure 34).

Figure 34 represents the probability that the player with *a* pegs beats the player with *b* pegs, denoted by *a-b*, where *a* is a number on the *x*-axis and *b* is a number on the *y*-axis. For example, in the case of 3-3, the first player will win the game with probability of 83%. In the case of 3-6, the winning probability for a player with 3 pegs is zero. This means that the player loses the game or the game is a draw. An interesting case is 9-9, which means that when all the nine pegs are filled in, the first player will win the game with high probability.

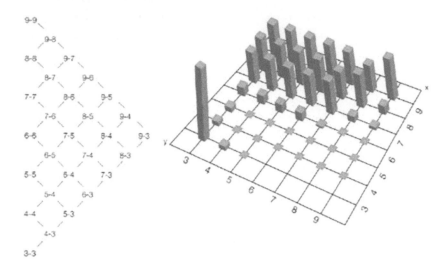

FIGURE 34 Pattern of pegs for *nine men's morris*, from Ralph Gasser (1996).

PROBLEMS

1. (*) Now is Black's turn in first step. Find a winning strategy for Black.

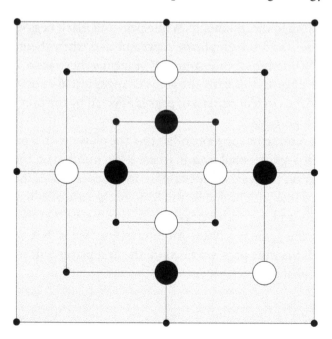

2. (*) Now is Black's turn in second step so that Black should push pegs. Show that White can always win the game.

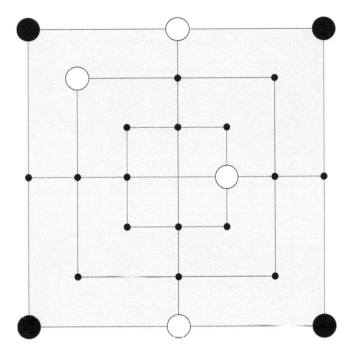

3. (**) According to Gasser's graph, when there are three Black pegs and five White pegs, there is a little chance that Black wins the game. Find such an example.

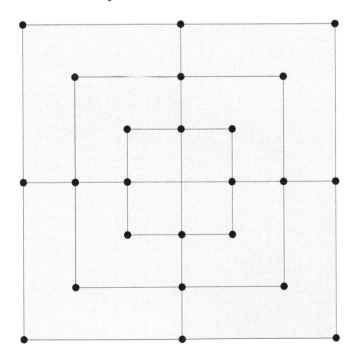

4. (***) What is the condition that always gives a win for three men's morris? Find this under first step and second step.

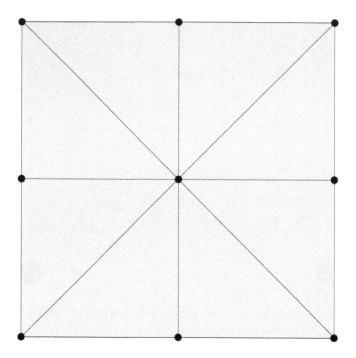

The Game of Quatrainment

Flip Neighbor Stones

THE GAME OF *QUATRAINMENT*

The game of *Quatrainment* is a game on a board of shape 4×4. Quatrain means a poem with four lines. *Quatrainment* was introduced by Sean Puckett, an American Software engineer, in 1984 in the magazine called <Compute!> and then became famous. Tom Gantner, a mathematician at the University of Dayton found a solution by representing it mathematically (Figure 35).

DOI: 10.1201/9781003268024-8

Quatrainment requires two boards of shape 4 × 4. One is a starting board and the other is an ending board. Some spots of each board are marked as X. The goal of the game is to convert the Xs' positions in the starting board into the Xs' positions in the ending board. Flipping X becomes the empty place and flipping the empty place becomes X.

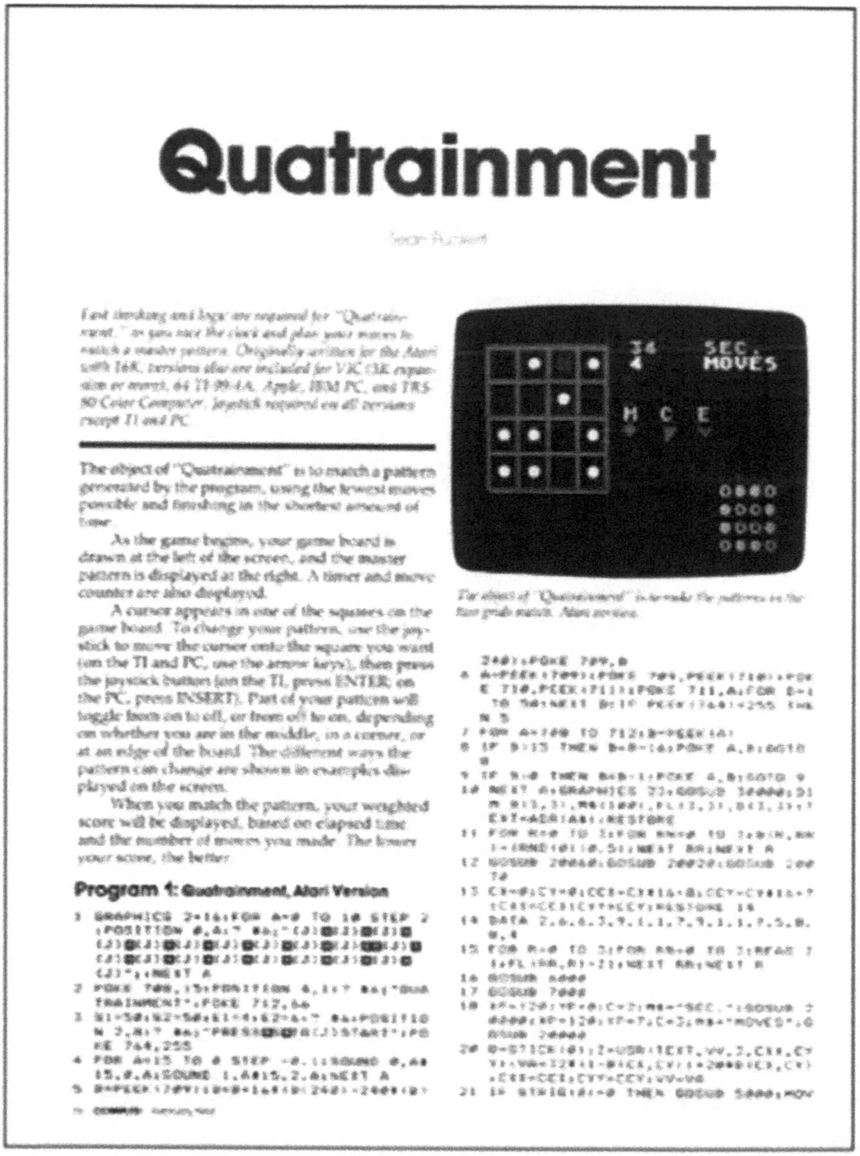

Quatrainment

The object of "Quatrainment" is to match a pattern generated by the program, using the fewest moves possible and finishing in the shortest amount of time.

As the game begins, your game board is drawn at the left of the screen, and the master pattern is displayed at the right. A timer and move counter are also displayed.

A cursor appears in one of the squares on the game board. To change your pattern, use the joystick to move the cursor onto the square you want (on the TI and PC, use the arrow keys), then press the joystick button (on the TI, press ENTER; on the PC, press INSERT). Part of your pattern will toggle from on to off, or from off to on, depending on whether you are in the middle, in a corner, or at an edge of the board. The different ways the pattern can change are shown in examples displayed on the screen.

When you match the pattern, your weighted score will be displayed, based on elapsed time and the number of moves you made. The lower your score, the better.

Program 1: Quatrainment, Atari Version

FIGURE 35 Sean Puckett's *quatrainment* in the journal Compute!

If one can flip any place without any condition, this game can be done easily. Here is a condition. When we flip X, some of its neighboring places should be also flipped. There are three cases depending on the position of X (see Figure 36 for details).

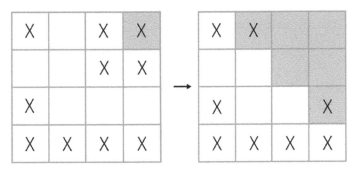

Rule 1: When choosing any corner position, that position and five neighboring positions are flipped.

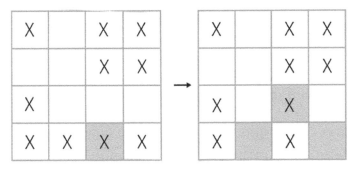

Rule 2: When choosing any position on an edge, three neighboring positions are flipped.

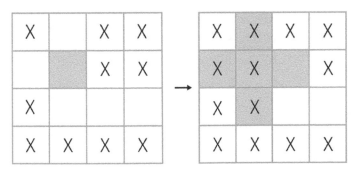

Rule 3: For the remaining positions, that position and four neighboring positions are flipped.

FIGURE 36   Rules for quatrainment.

MATHEMATICAL PRINCIPLE

Let us apply the above three rules in the 3×3 board. See Figure 37. We call it '*quatrainment of order 3*'. Rule 2 in this case requires that the chosen position needs to be flipped as well.

Here is the arithmetic we apply.

$$0+0=0$$

$$0+1=1$$

$$1+0=1$$

$$1+1=0$$

The reason why $1+1=0$ instead of $1+1=2$ is that if we think 1 as an odd number and 0 as an even number, then the fact that odd number+odd number=even number is represented as $1+1=0$.

We can represent the board with Xs as an 3×3 array with 0 and 1. If there is an X, then put "1" and if there is no X, then put "0". For example, the starting board and the ending board are transformed into 3×3 arrays as in Figure 38.

Flipping a corner position of a staring board is the same as adding two arrays in Figure 39.

Therefore, we make nine arrays consisting of 0 and 1 corresponding to the nine cases such as four corner positions, four edge positions, and the center position (see Figure 40).

Therefore, if we represent the starting board, the ending board, and the nine flipping rules as arrays of 0 and 1 and add them using the matrix addition rule, then we can find an explicit solution. A matrix means an

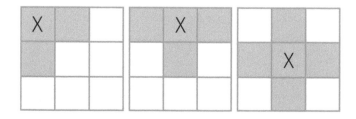

FIGURE 37   *Quatrainment of order 3.*

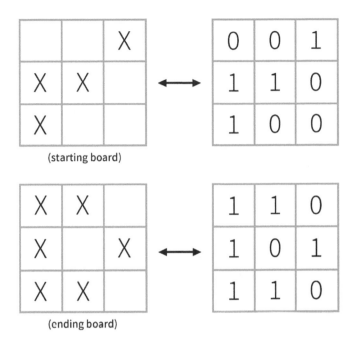

FIGURE 38   Mapping the board to 3×3 array with 0 and 1.

1	1	0
1	0	0
0	0	0

+

0	0	1
1	1	0
1	0	0

=

1	1	1
0	1	0
1	0	0

FIGURE 39   Sum of two boards.

array of numbers in general. When we add two matrices, we add the corresponding components. To add two matrices, the size of the matrices should be the same.

For convenience, assume that A is a staring array and B is an ending array. Flipping A is the same as adding some $M_1 \sim M_9$ repeatedly. If we apply the scalar multiple of a matrix A by a scalar c, then the whole process of the solution can be described as a matrix equation. If A is flipped using the array M_1 c_1 times, then we can represent it as $A + c_1 M_1$. Therefore, in order to get the ending board from the given starting board by several flips, there is a solution to the following matrix equation $A + c_1 M_1 + c_2 M_2 + \cdots + c_9 M_9 = B$ for some $c_1 \sim c_9$. Now this equation becomes the following system of linear equations.

M_1			M_2			M_3		
1	1	0	1	1	1	0	1	1
1	0	0	0	1	0	0	0	1
0	0	0	0	0	0	0	0	0

M_4			M_5			M_6		
1	0	0	0	1	0	0	0	1
1	1	0	1	1	1	0	1	1
1	0	0	0	1	0	0	0	1

M_7			M_8			M_9		
0	0	0	0	0	0	0	0	0
1	0	0	0	1	0	0	0	1
1	1	0	1	1	1	0	1	1

FIGURE 40   Representation of all the rules by arrays of 0 and 1.

c_1M_1	c_2M_2	c_3M_3	c_4M_4	c_5M_5	c_6M_6	c_7M_7	c_8M_8	c_9M_9	
$0 + c_1$	$+ c_2$	$+ 0$	$+ c_4$	$+ 0$	$+ 0$	$+ 0$	$+ 0$	$+ 0$	$= 1$
$0 + c_1$	$+ c_2$	$+ c_3$	$+ 0$	$+ c_5$	$+ 0$	$+ 0$	$+ 0$	$+ 0$	$= 1$
$1 + 0$	$+ c_2$	$+ c_3$	$+ 0$	$+ 0$	$+ c_6$	$+ 0$	$+ 0$	$+ 0$	$= 0$
$1 + c_1$	$+ 0$	$+ 0$	$+ c_4$	$+ c_5$	$+ 0$	$+ c_7$	$+ 0$	$+ 0$	$= 1$
$1 + 0$	$+ c_2$	$+ 0$	$+ c_4$	$+ c_5$	$+ c_6$	$+ 0$	$+ 0$	$+ 0$	$= 0$
$0 + 0$	$+ 0$	$+ c_3$	$+ 0$	$+ c_5$	$+ c_6$	$+ 0$	$+ 0$	$+ c_9$	$= 1$
$1 + 0$	$+ 0$	$+ 0$	$+ c_4$	$+ 0$	$+ 0$	$+ c_7$	$+ c_8$	$+ 0$	$= 1$
$0 + 0$	$+ 0$	$+ 0$	$+ 0$	$+ c_5$	$+ 0$	$+ c_7$	$+ c_8$	$+ c_9$	$= 1$
$0 + 0$	$+ 0$	$+ 0$	$+ 0$	$+ 0$	$+ 0$	$+ 0$	$+ c_8$	$+ c_9$	$= 0$

It seems very difficult to solve it by hand. Noting that adding any M_i twice results in the zero array, we may assume that $c_1 \sim c_9$ is 0 or 1. It is then easy to find the values of $c_1 \sim c_9$. First plug in 0 or 1 from c_1, c_2, c_3. There are eight cases. For each case, plug in 0 or 1 from c_1, c_2, c_3, c_4 to see if there is a solution. One can see that there is a solution

$c_1 = 0, c_2 = 0, c_3 = 1, c_4 = 1, c_5 = 0, c_6 = 0, c_7 = 1, c_8 = 0, c_9 = 0.$ This shows that which rules should be applied to arrive at the ending board. We do not need to be tied up with this formula. Just enjoy the game by trial and error.

PROBLEMS

1. (*) In the *quatrainment of order 3*, the staring board A and the ending board B are given as follows. How many and which flips are needed? Double check your answer with the formula in the chapter.

$$A = \begin{array}{|c|c|c|} \hline 0 & 1 & 0 \\ \hline 1 & 0 & 1 \\ \hline 0 & 0 & 0 \\ \hline \end{array} \qquad B = \begin{array}{|c|c|c|} \hline 1 & 0 & 1 \\ \hline 0 & 1 & 0 \\ \hline 1 & 0 & 0 \\ \hline \end{array}$$

2. (*) In the original *quatrainment*, the starting board A and the ending board B are given as follows. How many and which flips are needed?

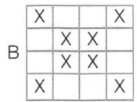

3. (**) Find a formula (a system of linear equations) for the game of *quatrainment* similar to the *quatrainment of order 3*.

4. (***) If you want the game of *quatrainment of order 5*, what kinds of rules are needed? Find a formula (a system of linear equations) for this game.

n-Queens Game and Puzzle

Playing Chess Only with Queens

THE n-QUEENS GAME WITH MANY QUEENS

There is a piece called a queen in chess game. The queen can move to any empty position horizontally and vertically just like a 'car' in the Korean chess game called 'Jang Gi'. It can also move diagonally. A queen is one of the most powerful pieces in Chess. Besides a queen, there are other pieces

DOI: 10.1201/9781003268024-9

including a king, a rook, a pawn, a knight, and a bishop. However, the *n-queens* game uses $n \times n$ chessboard and many queens. You may use go board and go stones (Figure 41).

The *n-queens* game started from the eight-queens puzzle where the eight queens are placed according to the rule on the 8×8 chess board. It was first introduced in 1848 by Max Bezzel, a German chess researcher. Two years later, a German doctor Franz Nauck figured out the solution of the eight-queens puzzle and extended it to the $n \times n$ chess board. Famous mathematicians including Carl Friedrich Gauss had an interest in the puzzle and studied its solutions. In 2016, Glen Van Brummelen at Bennington College and his student Hassan Noon converted the puzzle game as a two-player game.

Let us take a look at the rule of the *eight-queens* game. Prepare the 8×8 chess board. Make four white pieces and four black pieces, all of which represent queens. Decide who will play first and play alternately. The rule is that players should place a new queen which should not be attacked. If a player cannot put a queen, the player loses the game.

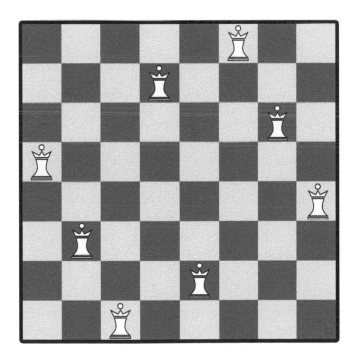

FIGURE 41   Queens which do not attack each other.

ADVANTAGE FOR THE FIRST PLAYER?

When you play the game several times, you will see that there would be at least five queens on the board to end the game. Since queens cannot be placed horizontally and vertically, there can be at most eight queens on the board. Therefore, the first player should win the game at the fifth or seventh turn. The second player should win the game on the sixth or eighth turn.

When $n = 4$, the first player can always win the game. For example, if the first player puts the queen at the corner, then the second player has two options. After this option, the first player has an empty place, and hence wins the game.

Corner position: Try non-corner positions.

Is there a winning strategy for other values of n? For example, when $n = 7$, the first player can always win the game (Figure 42). The first player should put the queen at the center (Figure 42a). The second player will put a queen like Figure 42b. Then the first player can put a queen in a symmetric way (Figure 42c). By continuing this way, the first player can always win the game. In general, this strategy works when n is odd. However, there is no winning strategy known when n is even.

MATHEMATICAL PRINCIPLE

Unlikely the *n-queens* game between two players, the *n*-queens puzzle is to place *n queens* on the $n \times n$ board with non-attacking positions. Depending on n, there can be no way or many ways to put the *n queens*. If $n = 27$, there are 234, 907, 967, 154, 122, 528 ways.

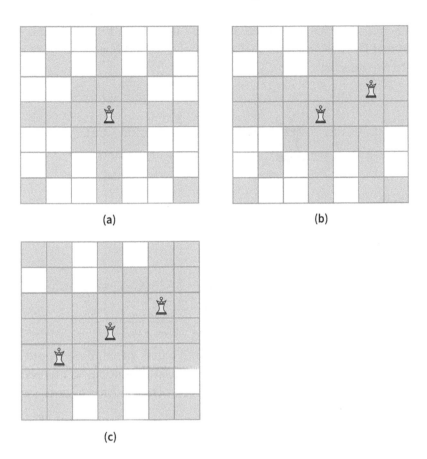

FIGURE 42 $n = 7$ case, queens' move.

There is at least one way for any n, except for $n = 2$, $n = 3$. We just know there is a way but we do not know an explicit formula for the number of ways. When n increases infinitely, we do not know whether the number diverges or converges to a number. When n is 1–27, there is an exact number of ways.

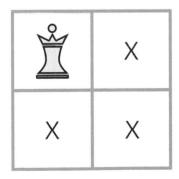

When $n = 2$, put a first queen at the left top corner first. Then it is easy to see that there is no available place to put a second queen. So there is no way to two queens.

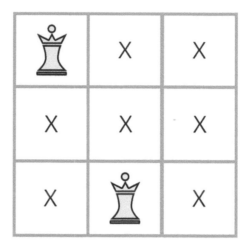

When $n = 3$, there are three places to put a first queen. First we put it at one of the four corners. Then there are two available places. If one place is chosen, then the other place is not available. Hence three queens cannot be on the 3×3 board.

When a first queen is placed on one of the edges, it is easy to see that we cannot put three queens together. When a first queen is placed on the center, it attacks all positions. Therefore, when $n = 3$, we cannot put three queens.

When $n = 4$, there is one way to put four queens as follows. By shifting this shape with respect to the x-axis or the y-axis, we can obtain another solution so that we have two solutions at the end. When we want to find all solutions, we first find a solution and then apply the symmetry of the solution.

It was a math genius Gauss who found the number of ways to put the eight queens for the 8-queens puzzle. In 1850 when there was no computer, Gauss found that there are exactly 92 ways in the 8-queens puzzle. Since the 8-queens puzzle has 64 spots, there are $\binom{64}{8} = 4{,}426{,}165{,}368.$ Hence using the rule that if one queen is placed then the rows and columns containing the queen cannot be selected, we can reduce the possibility. Of course, we can consider the diagonal rule but it already reduced many possibilities.

Since the chess board also consists of the eight columns, we choose any spot in the first column, and choose another spot among the seven available spots in the second column. Continuing this way, we have at most $8! = 40{,}320$ possible ways to place 8 queens. In 2002, Zongyan Qui from Peking University showed that there are 12 ways up to symmetry by using a computer search as in Figure 43.

We rotate each of the first 11 solutions (Figure 43) 90 degrees to obtain 3 more solutions and flip each horizontally, vertically, and diagonally to obtain 4 more. These imply that there are $11 \times 8 = 88$ solutions. The 12th solution of Figure 43 results in 3 more solutions after rotation and reflection. Therefore, there are exactly 92 distinct solutions for the 8-queens puzzle.

It has been known up to now that there are exact solutions for the n-queens puzzle when n is less than or equal to 27.

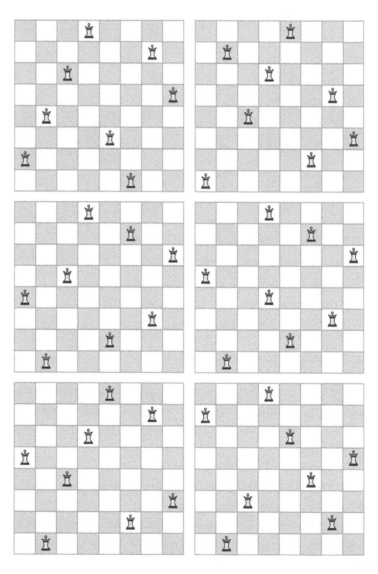

FIGURE 43   Eight-queens puzzle solutions. Wikipedia.

(Continued)

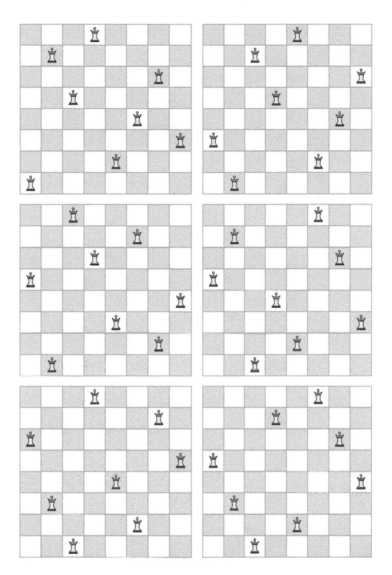

FIGURE 43 (*Continued*)   **Eight-queens puzzle solutions. Wikipedia.**

PROBLEMS

1. (*) Play the *8-queens* game with friends ten times. Compute the average number of turns to win or lose the game. Explain why there should be at least five turns to win the game.

2. (**) Play the *6-queens* game with friends ten times and discuss a winning strategy.

3. (*) In the *4-queens* game, find all symmetric positions of the given 4 queens below.

X	X	♛	X
♛	X	X	X
X	X	X	♛
X	♛	X	X

4. (**) Up to the symmetry, show that there are only two solutions to place 5 queens in the *5-queen* game. Then find the ten symmetric solutions.

5. (***) Up to the symmetry, show that there is only one solution to place 6 queens in the *6-queens* game. Then find the four symmetric solutions. In particular, this number is smaller than the case of the *5-queens* game.

REFERENCE

Qiu, Zongyan (February 2002). "Bit-vector encoding of n-queen problem". *ACM SIGPLAN Notices*. **37** (2): 68–70.

CHAPTER 10

Light Out

Turn Off Lights by Linear Equations

LIGHT OUT BY PUSHING A BUTTON

In 1978, a game designer Bob Doyle from NASA made a game machine called 'Merlin'. *Merlin* was so popular that two millions of *Merlin* were sold in the USA in 1980. *Merlin* contains six games including *Tic-Tac-Toe*. Among them, '*Magic Square*' is the most interesting game mathematically. The *Magic Square* is the original game for the *light out game*.

Magic Square is played on a 3×3 array where each entry has a button so that one can turn on or off the light until there remains only one light

DOI: 10.1201/9781003268024-10

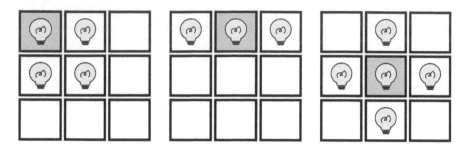

FIGURE 44   Three types of rules (i), (ii), (iii) from the left.

at the center. If one pushes a button, then the neighboring lights are on or out, which makes the game not easy. For example, if one pushes a button at a corner, then the button, the two buttons next to it and its diagonal button change the status of the light (Figure 44(i)). If one pushes a button on an edge, the button and the two buttons next to it change the status of light (Figure 44(ii)). If one pushes the center button, the button and its four neighboring buttons change the status of the light (Figure 44(iii)).

In 1987, a Canadian mathematician Don Pelletier introduced a strategy for doing well *Magic Square* in the American Mathematical Monthly. He used vectors to represent the state of on and out. Let us get into the detail.

First, we denote the light on by 1 and the light out by 0. Read the top line from the left to the right to get an initial binary vector of length 9. Thus, all the rules can be represented by the nine binary vectors. If the initial vector is the zero vector, when the button 3 is pressed, the zero vector becomes $(0,1,1,0,1,1,0,0,0)$. By adding these kinds of vectors, we need to make $(1,1,1,1,0,1,1,1,1)$ which corresponds to (on, on, on, on, off, on, on, on, on) in order to win the game. Now, 'adding' here is a little different from adding two binary numbers. For example, $0+0=0$, $1+0=0+1=1$ as usual, but $1+1=0$. This means that there is no carry in this addition (Figure 45).

1	2	3
4	5	6
7	8	9

1	2	3
4	5	6
7	8	9

1	2	3
4	5	6
7	8	9

FIGURE 45   Representation of each status by a binary vector as $(0,0,0,0,0,0,0)$, $(0,1,1,0,1,1,0,0,0)$, $(1,1,1,1,0,1,1,1,1)$ respectively.

LIGHT OUT

Light out was made by an American game company Tiger Electronics in 1995. There are on/out buttons on the 5×5 array. If you press a button, then the button and its neighboring buttons switch the status of the light. Unlike *Magic Square*, the diagonal buttons do not change. The goal of the light out is to turn off all the lights (Figure 46).

Mini light out is a shortened light out based on the 4×4 array. While light out switches the status of the buttons next to a given button, mini light out also switches the status of the buttons opposite to a given side button because we assume that a top side button and a corresponding bottom side button are neighbors and a left side button and a corresponding right side button are neighbors.

In Mathematics, we call such a structure a torus. A torus appears often in Topology. A torus is topologically equivalent to a square. This is because we can attach opposite sides to make a torus. Therefore, pressing one

FIGURE 46 **Lights out.**

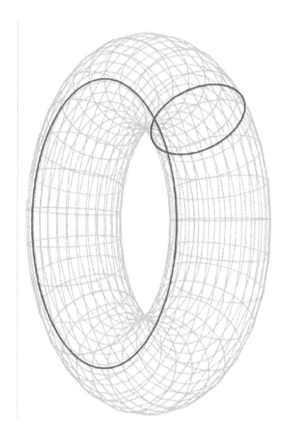

FIGURE 47   Torus (CC0.1).

button changes the status of the neighboring lights, which is the difference with the *light out game*. Just like the *light out game*, the aim of this game is to turn off all the lights given arbitrarily lit lights (Figure 47).

Both *magic squares* and *light out game* can be solved by setting a system of linear equations. For simplicity, we consider 2×2 mini *light out game*. The rule of this game is the same as the *light out game*. Pressing one button changes the status of the lights in the button itself and its two neighboring buttons. If we think of "on" as 1 and "off" as 0, then there are four cases depending on the position of the button.

$$
\begin{array}{cc}
0 & 0 \\
0 & 0
\end{array}
\rightarrow
\begin{array}{cc}
\boxed{1} & 1 \\
1 & 0
\end{array}
\text{ or }
\begin{array}{cc}
1 & \boxed{1} \\
0 & 1
\end{array}
\text{ or }
\begin{array}{cc}
0 & 1 \\
1 & \boxed{1}
\end{array}
\text{ or }
\begin{array}{cc}
1 & 0 \\
\boxed{1} & 1
\end{array}
$$

Suppose that there are two lights on as follows. Turn off all the lights by pressing the least number of buttons.

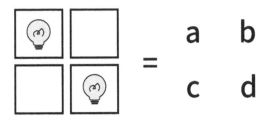

After some trials and errors, you can make all status 0s. In fact, two buttons will be enough.

$$\text{(image)} = \begin{array}{cc} a & b \\ c & d \end{array}$$

Each button can correspond to one of letters, a, b, c, and d. Remember that $0+0=0$, $1+0=0+1=1$, $1+1=0$. Since a is lit and b and c are not lit, we have ① $a+b+c=1$ by this arithmetic. Since a and d are lit, ② $a+b+d=0$. In a similar manner, ③ $a+c+d=0$, ④ $b+c+d=1$. Now we can solve the system of these equations. By adding ① to ②, we get ⑤ $c+d=1$. If we press one button twice, we have $0+0=0$, $1+1=0$ because it has the effect that we have done nothing. In other words, we have $2a=0$, $2b=0$.

If we subtract ⑤ from ③, we get $a=1$. Subtracting ⑤ from ④ results in $b=0$. Plug $a=1$, $b=0$ into ① to get $c=0$ and $a=1$, $b=0$ into ② to get $d=1$. Therefore, press buttons a and d.

Besides the square, one can play *light out game* on various figures or graphs. Choose your favorite figure to make your own *light out game*.

PROBLEMS

1. (*) Let us enjoy the 2×2 mini *light out game*, by pressing the minimum number of buttons to solve the following problems. Solve it by your intuition first and by a system of equations.

    (a) 1 1   (b) 1 0   (c) 1 1
         0 0   0 0   1 1

2. (**) In *magic square* game, explain why the center light can be off at the end regardless of any initial status.

3. (**) In the *light out game*, we are given the following. What is the minimum number of buttons to turn off all the lights?

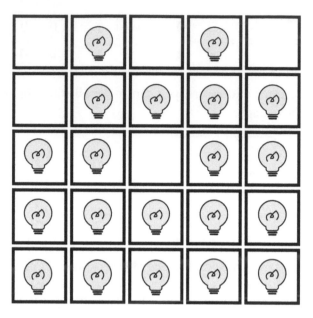

4. (***) In the mini *light out game*, explain why all the lights can be off at the end regardless of any initial status.

CHAPTER 11

1258 Board Game

Magic Square and Orthogonal Latin Square

THE 36 OFFICERS PROBLEMS

In 1782, Leonhard Euler proposed the following 36 officers' problem.

Is it possible to arrange a delegation of six regiments, each of which sends a colonel, a lieutenant-colonel, a major, a captain, a lieutenant, and a sub-lieutenant in a 6 by 6 array such that no row or column duplicates a rank or a regiment?

DOI: 10.1201/9781003268024-11

Euler tried many ways and conjectured that such an array is not possible. If n is 5, that is, the problem to arrange five regiments with five different ranks without duplications in a 5 by 5 array is possible. However, if n is 6, the solution is hard to find. This problem was not solved for 120 years.

Finally, in 1901, a French mathematician Gaston Tarry proved that such an array is not possible by enumerating several thousands of cases. In 1984, a Canadian mathematician Doug Stinson proved it by using a combinatorial design theory and in 1994, an American mathematician Steven Dougherty reproved it by using Coding Theory and Finite Geometry. Their proofs are based on advanced mathematics. There is no known proof based on an elementary method which can be understood by a college student.

Let us get back to Euler's conjecture. Euler also conjectured that it will not be possible to arrange n regiments with n different ranks without duplications in an n by n array when $n = 4k + 2$ $(k \geq 2)$. However, in 1959, Raj Chandra Bose, Sharadchandra Shankar Shrikhande, and Ernest Parker disproved this conjecture when $n = 10$. Since then, the conjecture was disproved for any $n = 4k + 2$ $(k \geq 2)$.

SEOK-JEONG CHOI WHO STUDIED ORTHOGONAL LATIN SQUARES

The arrays appearing in Euler's conjecture are called orthogonal Latin squares. We recall that the square array with numbers 1 to n^2 where the row sum, the column sum, and the diagonal sum are all the same is called a magic square. Latin square is a variant of a magic square such that each row has n numbers (symbols) such that each number (symbol) appears exactly once and the same is true for each column. *Sudoku* is a good example of a Latin square. When two Latin squares of an order n (meaning size of $n \times n$) are superimposed and the ordered paired entries in the positions are all distinct, we call such a pair of Latin squares *orthogonal*.

The study of orthogonal Latin squares is regarded as the origin of Combinatorics. In Europe, Euler wrote a paper on orthogonal Latin squares in 1776. However, it was discovered that Seok-Jeong Choi from Korea studied orthogonal Latin squares 61 years earlier than Euler. He was a prime minister in Chosun dynasty and introduced a pair of orthogonal Latin squares of order 9 for the first time in his book Koo-Soo-Ryak (or Gusuryak). Using this, he also constructed a magic square of order 9. It is unknown how and why he discovered such a pair of orthogonal Latin squares of order 9 (Figure 48).

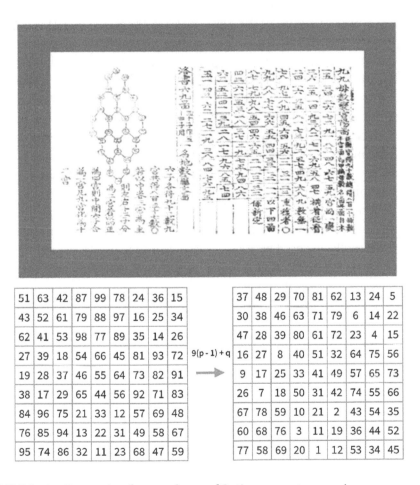

FIGURE 48 Conversion from orthogonal Latin squares to a magic square.

1258 GAME AND ORTHOGONAL LATIN SQUARES

In this section, I would like to introduce a board game called *1258 game* which I invented in 2017. What is the connection between *1258 game* and orthogonal Latin squares? Simply speaking, *1258 game* is to find values in the row or column of a pair of orthogonal Latin squares of order 4 consisting of 1, 2, 5, and 8. Let us look at the rules of *1258 game*.

First determine the order of the players, shuffle the cards, put nine cards on the deck in the form of a 3 by 3 array. The remaining cards are placed on the center card of the deck. In your turn, pick one card from one of the eight cards or one from the pile of the center card. If one card is picked from the eight cards, move one card from the pile to the empty spot. Each player repeats this process. If there are at least four cards in your hand, check if the ones places and the tens places in the four cards consist of 1, 2, 5, 8 in any order. Because of the symmetries in the numbers 1, 2, 5, 8, you may rotate the cards 180 degrees or flip them. If you find such four cards, call 'Ola' and put them aside. Ola has three types of scores. If all colors of the four cards are the same, 4 scores are given. If all colors are different, 2 scores are given. For other cases, 1 score is given. Play this game until there is no more card left in the deck. Then add all the scores each player obtained. Rank the players based on the scores. If two total scores are equal, the player with more 4 scores will be ranked higher. If the 4 scores are also equal, the player with more 2 scores will be ranked higher. If the 2 scores are also equal, the game is a tie.

In Mathematics, a Latin square is represented as a matrix of order n. A matrix of order n is an $n \times n$ square array whose entries are numbers or letters, normally enclosed by a parenthesis or a bracket. Therefore, we can make a Latin square of order 4 based on 1, 2, 5, 8, and represent them as the following matrices A and B.

Now we superimpose A and B to get a matrix C.

$$A = \begin{pmatrix} 1 & 2 & 5 & 8 \\ 5 & 8 & 1 & 2 \\ 8 & 5 & 2 & 1 \\ 2 & 1 & 8 & 5 \end{pmatrix} \quad B = \begin{pmatrix} 1 & 2 & 5 & 8 \\ 8 & 5 & 2 & 1 \\ 2 & 1 & 8 & 5 \\ 5 & 8 & 1 & 2 \end{pmatrix} \Rightarrow C = \begin{pmatrix} 11 & 22 & 55 & 88 \\ 58 & 85 & 12 & 21 \\ 82 & 51 & 28 & 15 \\ 25 & 18 & 81 & 52 \end{pmatrix}$$

If you look at each row or column of C, the ones places contain 1, 2, 5, 8 exactly once and the tens places contain 1, 2, 5, 8 exactly once. This is true for the two diagonals. In other words, if we regard each entry of the matrix

C as a card with two digits, then each row or column of *C* corresponds to the four cards called 'Ola'. From now on, enjoy the *1258 game* and try to find any mathematical properties of *1258 game*.

If one card is picked from the eight cards, move one card from the center to the empty spot.

If one card is picked from the center, no more action is needed.

THE RULE OF THE *1258 GAME*

1258 cards consist of 96 cards and two joker cards; 16 cards have two digits from 1, 2, 5, 8, and there are 6 distinct colors so that there are $16 \times 6 = 96$ cards. Any card can be rotated 180 degrees and can be also flipped to get a possibly new number. For example, 15 can be transformed into 51, 12, 21.

The following describes some patterns of the numbers:

1. When rotated or flipped, the numbers are the same: 11 88.

2. When rotated or flipped, the numbers produce another numbers: 18, 81, 22, 55.

3. When rotated or flipped, the numbers produce three new numbers: 12, 21, 15, 51, 28, 82, 58, 85.

PROBLEMS

1 (*) The sum of all the numbers in Ola is always 176. Why is it so?

$$12 + 21 + 58 + 85 = 176$$

2 (*) The following matrix C becomes a magic square of order 4 such that the sum of the diagonal is 176.

$$C \;=\; \begin{pmatrix} 11 & 22 & 55 & 88 \\ 58 & 85 & 12 & 21 \\ 82 & 51 & 28 & 15 \\ 25 & 18 & 81 & 52 \end{pmatrix}$$

3 (*) Using matrix C, make a magic square of order 4 whose row sum, column sum, and diagonal sum is 68.

4 (***) Find other examples like C in Problem 2.

5 (**) We can represent each Ola card as a bijective function between {1, 2, 5, 8} and {1, 2, 5, 8}. Why is that so?

6 (**) How many Ola cards of the circulant form like 12, 25, 58, 81 do there exist? That is, the ones place is connected with the tens place.

   (Hint: Such an Ola card has the property that $f(a)=a$ is not true for any value of a.)

CHAPTER 12

Switching Game

Shannon's Network Game

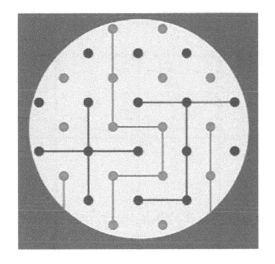

FIND A DETOUR

The *Shannon switching* game was invented in 1951 by Claude Shannon, who is an American mathematician and engineer well known as the father of information theory. When there was a disconnection in a network connecting many computers, he was trying to find a detour connecting the network. Motivated by this situation, *switching* game was made. Shannon studied how to send information such as letters, sound, image reliably through a noisy channel. One of his papers titled as "The Mathematical

DOI: 10.1201/9781003268024-12

Theory of Communication" is a foundational paper for information theory and is a classic in communication theory.

The *Shannon switching* game (simply *switching* game) is usually played on a square lattice graph. Recall that a graph consists of vertices and edges. An edge is a line connecting two vertices. First, determine two vertices on the graph as a starting point and an ending point. One player is called 'Short' who colors edges, and the other player is called 'Cut' who deletes edges. The goal of Short is to make a path connecting the starting point and the ending point. The goal of Cut is to interrupt Short not to make such a path.

Let us take a look at *switching* game on a square lattice graph with 25 vertices. Short(green color) colors edges in order to make a path from vertex A to vertex B, and Cut(red color) removes edges in order for Short not to make a path connecting A and B. In the graph below, Cut removes the 16th edge so that Short cannot arrive at B. Therefore, Cut wins the game.

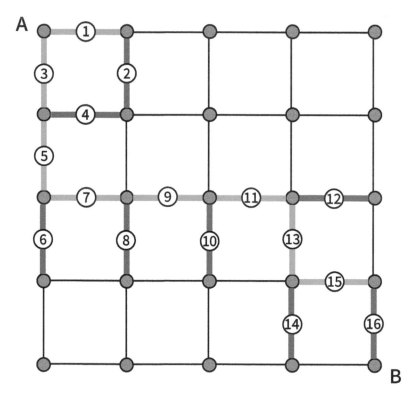

In the following graph, even though Cut has started first and Short has followed, Short can always win the game.

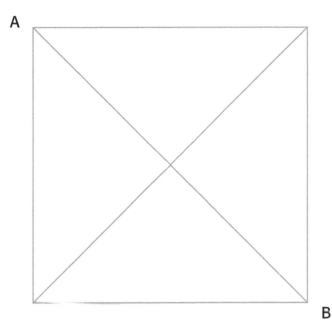

In 1964, a mathematician Alfred Lehman proved that there is a way to win *switching* game using a very complicated method. In 1996, Richard Mansfield reproved Lehman's result using a less complicated method. However, there is no explicit solution for *switching* game. Please try by yourself!

In spite of different look, '*Bridg-It*' and '*Hex*' are played in a similar way (Figure 49).

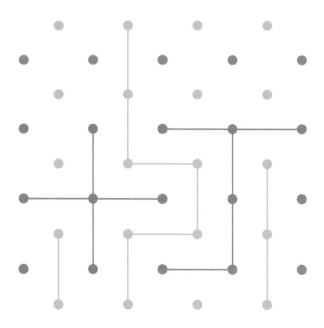

FIGURE 49   Bridg-It game, Green wins.

BRIDG-IT

Bridg-It was made by David Gale who also made *Chomp*. Unlike *switching* game, *Bridg-It* uses a rectangular lattice without edges where there are n vertices vertically and $n + 1$ vertices horizontally. Rotate another rectangular lattice of the same size 90 degrees to superimpose it with the original one to play the game. See the picture of Bridg-It.

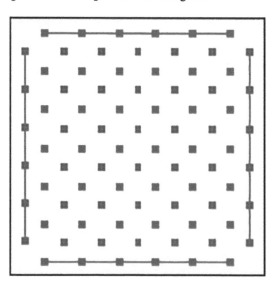

The rule of *Bridg-It* is as follows. A player who plays on the original lattice (blue color in the picture) is called 'Right' and a player who plays on the rotated lattice (red color in the picture) is called 'Left'. Left chooses one of the *n* vertices on the top as a starting point and connects vertices in the rotated lattice until Left arrives at any vertex at the bottom. Similarly, Right starts from any vertex on the left side and tries to reach any vertex on the right side.

Left and Right draw lines connecting adjacent vertices in their own lattice and should not cross their lines. The person who reaches the other end wins the game. If we separate the two lattices, then the two players are playing *switching* game at the same time. The condition that Left and Right should not cross their lines is equivalent to the condition that Cut in *switching* game removes edges.

HEXAGONAL GAME, *HEX*

Hex is a two-player board game on an 11×11 rhombus-shape board consisting of *hex*agons. Just like *Bridg-It*, the player who connects *hex*agons first in the opposite sides wins the game. The only difference is that each player puts his/her stones instead of connecting lines.

Hex was invented by a mathematician and poet Piet Hein in 1942 and was claimed to be rediscovered independently and popularized by John Nash in 1948 who is well known for his game theory. John Nash is the only person to be awarded both the 1994 Nobel prize in Economic Sciences and the 2015 Abel prize. Nash proved that the first player can always win the game and also proved the *Hex* theorem which states that *Hex* cannot end in a draw. In 1952, Parker Brothers, an American toy and game manufacturer, commercialized it. This game is still popular in the USA and Europe (Figure 50).

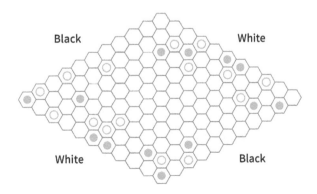

FIGURE 50   Hex game.

According to the *Hex* theorem, you can win the game by interrupting the other player's stones instead of connecting your own stones. Since there is no draw, if one player cannot connect his/her opposite sides, then the other player will win the game.

Let us look at the game in more detail. Each *hex*agon in the board can be regarded as a vertex. Two players are called White and Black. White starts from any position on one edge (marked green in the picture) and puts white stones to connect to any position on the opposite edge while interrupting black stones. Likewise, Black tries to connect black stones while interrupting white stones. In other words, the *Shannon switching* game is played by White and Black at the same time. *Hex* is usually played on a 11×11 rhombus-shape board but the board size can be variously adjusted.

Who can win the 3×3 *Hex* game? A player who puts on the center can win the game. As in Figure 51, if there is a black stone in the center, Black can make a path connecting two opposite edges wherever White puts stones. It is more complicated for the 4×4 *Hex* game. The first player who puts one stone in the positions 1, 2, 3, or 4 can win the game. Otherwise, lose the game. Then as the board size increases, where should we put stones to win the game? No explicit method is known. However, as indicated in the beginning, Nash proved that the first player can always win the game.

A surprising thing about *Hex* game is that there is no draw. This is called the *Hex* theorem. As we recall, tic-tac-toe game has a tie often. Hence this theorem is not obvious. In 1979, Gale proved that the *Hex* theorem and the Brouwer fixed point theorem are equivalent (Figure 52).

The Brouwer fixed point theorem named after Dutch mathematician Luitzen Brouwer can be a little difficult to explain mathematically. We explain it in a figurative way. Suppose that each of two persons go up or go down a mountain road with an angle of 45 degrees. One person goes

FIGURE 51   3×3 Hex and 4×4 Hex.

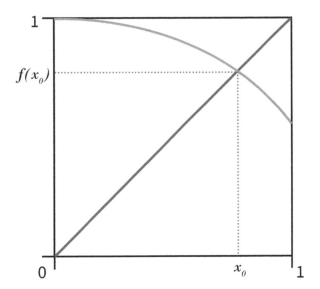

FIGURE 52   Fixed point theorem.

up from the bottom to the top of the mountain with a constant speed for an hour. The other person goes down from the top of the mountain to the bottom with any speed. Then the two persons should meet at some point. In Figure 53, x-axis denotes time and y-axis denotes height. The line graph $y = x$ and the curve $y = f(x)$ should always meet at $(x_0, f(x_0))$ because the curve $y = f(x)$ represents the height of the person going down and so this curve is continuous. Roughly speaking, this is the fixed-point theorem. Nash applied the fixed-point theorem to game theory to prove the Nash equilibrium which is the most common way to define the solution of a non-cooperative game involving two or more players.

PROBLEMS

1. (*) We can play *switching* game on Petersen's graph which contains ten vertices and 15 edges. It is a small graph that serves as a good example or a counterexample for many problems. Choose any two non-adjacent vertices outside as a starting point and an ending point. Can Short win the game by playing first or second?

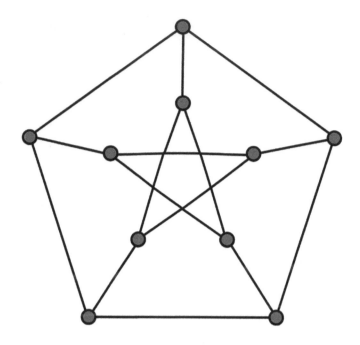

2. (*) Play *Bridg-It* on a 6×7 board with friends.

3. (***) Below is the *Hex* game with 3, 4, or 5 *hex*agons on each edge. Assuming White's turn (green stones), where are the black stones placed to win each game?

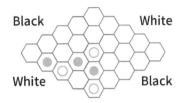

CHAPTER 13

Dots and Boxes

Occupy More Boxes

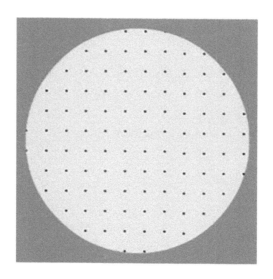

DOTS AND BOXES

Dots and Boxes is a pencil-and-paper game for two players. French mathematician Édouard Lucas in the 19th century described it and called it La Pipopipette. It is also known as the Dots Game, Boxes, Dots and Dashes, and Pigs in a Pen. Lucas studied the Fibonacci sequence and created the Lucas sequence by changing the first two initial values. Being interested a lot in recreational mathematics, he invented both *Dots and Boxes* and

DOI: 10.1201/9781003268024-13

the *Tower of Hanoi*. The *Tower of Hanoi* is a well-known game moving the disks of various size stacked in one rod to another rod via a third rod under certain rules. Lucas explained *Dots and Boxes* in a book on recreational mathematics in 1889.

Dots and Boxes is a game played on a square lattice of size 3×3, 4×4, 5×5, etc. As the name indicates, each player plays alternately and connects any two adjacent vertices (that is, makes an edge) to make a square box. It is an easy game with the following rules.

Each edge should be drawn one by one either horizontally or vertically. You do not need to draw all the edges of a square by your own drawing. Even if your opponent draws three edges of a square, you can take the square if you draw the last edge. The last rule is most important. If you make a square in your turn, you have to draw another edge. If you make another square by drawing an edge, you can continue to draw an edge. The game is over when all the edges are drawn. A winner is a person with more boxes. If there is an equal number of boxes, the first player loses the game.

Then let us play 3×3 *Dots and Boxes*. Let us call the first player Red who draws a red edge and the second player Blue who draws a blue edge. The box Red occupies is marked as 'R' and the box Blue occupies is marked as 'B'. In this way, we have distinguished the edges by colors and the boxes by 'R' or 'B' (Figure 53).

UC Berkeley mathematician Elwyn Berlekamp was interested in *Dots and Boxes* even when he was an elementary school student in 1946. In 1982, he co-authored the book "Winning Ways for your Mathematical Plays" with John Conway and Richard Guy and introduced a strategy for *Dots and Boxes*. In 2000, he wrote the book "*The Dots-and-Boxes Game*". Berlekamp did not find a winning strategy. In fact, finding a winning strategy for *Dots and Boxes* belongs to NP-hard, which is one of the most difficult problems in math and computer science. Nevertheless, he found a winning way for some cases.

The next strategy I assumes that Red plays first on the 4×4 lattice board. Since there is no occupied box and there is an even number of lines, it is Red's turn.

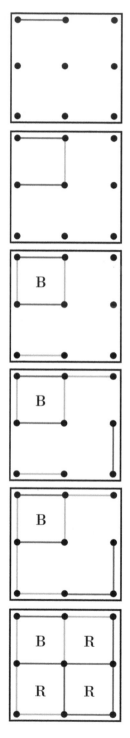

FIGURE 53 Example of *dots and boxes.*

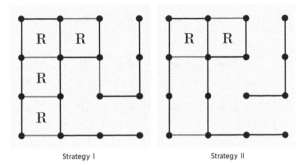

Strategy I   Strategy II

If Red draws one edge, then four boxes are marked as 'R' according to Strategy I, which is a common way. In this case, although there are four 'R's, there are still unmarked five boxes.

If Red chooses strategy I, Red should draw another edge. Since the result will be the same, we suppose that a dotted red line is drawn at the lower right side as in Figure 54 ①, ②, ③. Then Blue completes a square at the right bottom first and then completes four more boxes. Suddenly, since the score 4–0 changed to 4–5, Red loses the game.

If Red chooses strategy II, then Blue occupies two boxes as in Figure 54 ①, ②, ③ and draws an additional edge somewhere, say, the dotted line in the middle. Red makes use of the blue lines to occupy five boxes as in Figure 54 ①, ②, ③. In the beginning of strategy II, Red occupies less squares in order to throw the bait. Red obtains five more boxes so that Red wins the game with 7–2. This strategy is called a 'double-cross strategy'.

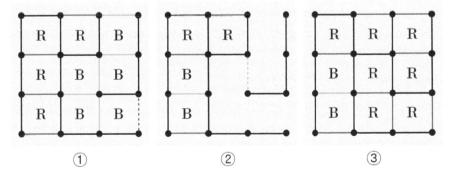

①   ②   ③

FIGURE 54   Which strategy is needed?.

FIGURE 55   4 × 4 board.

WINNING BY A CHAIN OF BOXES

Can the double-cross strategy work always? If the opponent also uses the double-cross strategy, how can you win the game? In order to apply the double-cross strategy, we need a condition that there is a chain of boxes in the lattice. In the previous example, Red's double-cross strategy was successful because after Blue draws an edge, Red can make five boxes in a successive way. If we can make at least n boxes successively after drawing one edge, we call such a shape 'a chain of boxes of length n'. Here we assume that n is at least 3.

A possible winning strategy for *Dots and Boxes* is as follows. Each player tries to force the opponent to make the first long chain because this chain of boxes is in fact a sacrifice to produce more boxes later.

For example, in *Dots and Boxes* on the 4 × 4 board (Figure 55), try not to make boxes in the beginning. Try to avoid the double-cross strategy if possible. Since this strategy is relative, it is hard to guess who will win the game at the end.

VARIOUS *DOTS AND BOXES* GAMES

There are *Dots and Boxes* games based on triangles and hexagons. For example, there are 28 vertices and 36 edges on the triangular graph board. A box needs four edges but a triangle needs only three edges. Hence, many triangles are formed easily. Even though the shapes are different, players can apply a double-cross strategy to make a long chain of boxes by yielding a short chain of boxes (Figure 56).

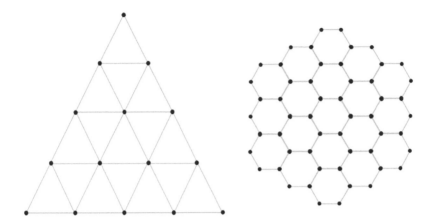

FIGURE 56   Dots and Boxes games based on triangles and hexagons.

STRINGS AND COINS GAME

There is a game called 'Strings and Coins' which is basically the same game as Dots and Boxes. Display coins in a square lattice which correspond to the boxes in Dots and Boxes. Connect coins with strings in four directions so that coins in the edges or corners will have 1 or 2 additional strings. Players cut the strings with a scissors alternately. If a player cuts the four strings, the player can take the coin and cut another string just like Dots and Boxes. A box in Dots and Boxes corresponds to a coin in Strings and Coins and the adjacent boxes correspond to the string connecting two coins. The edge sharing two boxes correspond to edge cutting. As in Figure 57, two boxes marked A means that two coins are not connected by any strings.

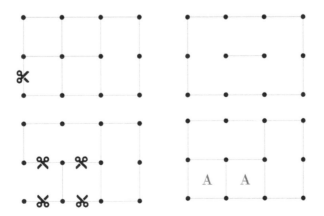

FIGURE 57   Rules for strings and coins.

PROBLEMS

1. (*) In *Dots and Boxes* on 3×3 lattice, one can see that who can win the game when there are 9, 10, or 11 edges drawn on the board. After playing several games, find which patterns exist.

2. (*) In the following *Dots and Boxes* on 3×3 lattice, which edge is the worst decision?

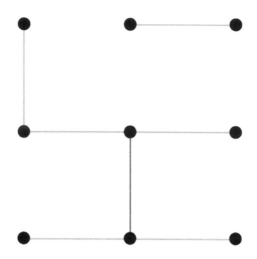

3. (***) In the following two *Dots and Boxes* on 4×4 lattice, you can win the game by drawing only one edge in each game. Where do you draw an edge?

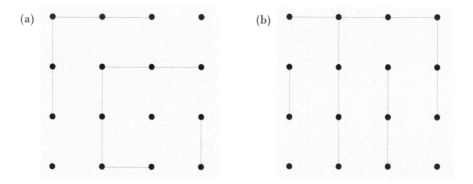

4. (****) Is there a winning strategy in *Dots and Boxes* on 4×4 lattice? Write down your own thought.

CHAPTER 14

Matricking

Factorization and Cube Net Game

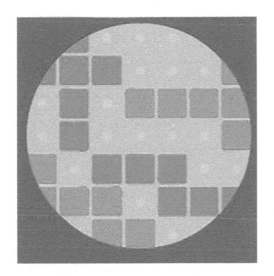

MATRICKING

Even a person who likes Mathematics can be nervous when the person learns a new mathematical concept. This is true especially when the formula is new and problems are not understandable. If one can learn some games related to mathematical concepts, learning mathematics can be enjoyable. Therefore, I have made a website called Matricking.

DOI: 10.1201/9781003268024-14

Matricking is a combination of Math, Trick, and King. Like an appetizer, Matricking can arouse mathematical curiosity and knowledge. It is based on simple rules so that one can learn Mathematics with fun. Matricking has several online games including *factorization*, *cube net*, *ratio*, and *acute triangle* (www.matricking.com). Each game is played on a square board with various levels against a computer. In a battle mode, 2~4 people can play together.

We explain each game with two players. The first player is denoted by Red and the second player is denoted by Blue.

FACTORIZATION GAME

Just as we write 12 as $2 \times 6, 3 \times 4$, factorization is a process to factor an integer into a product of two integers. Depending on an integer, there can be various ways to factor the integer. *Factorization* game represents each factorization as a rectangle so that the concept can be easily visualized.

For example, factorization with number 4 has the following five types of rules.

1. Since 4 is represented as 1×4, place four stones horizontally.

2. Since 4 is represented as 4×1, place four stones vertically.

3. Since 4 is represented as 2×2, place two stones horizontally and two stones vertically.

4. Place four stones diagonally. There are two ways.

In general, when n is given, one represents $n = ab$ so that there are a stones horizontally and b stones vertically. Of course, we allow two diagonals in order to have more fun. Each player puts stones alternately based on the rules. The person who can no longer play will lose the game. Do it on Figure 58.

If n is a prime greater than 2, the *factorization* game based on $n \times n$ board has a winning strategy for the first player. The first player places n stones in an odd number of turns. This means that the second player should places n stones in the same direction as the first player since that is the only way to place them. Since there are n turns and n is odd, the first player will win the game.

For example, in the board of size 3×3, as there is only one way to place three stones, the first player will always win the game. In general, if n is not

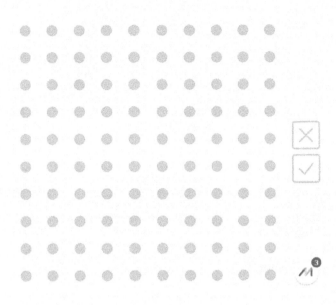

FIGURE 58   Board for the factorization game.

a prime number and the board size is $m \times m$ (m>n), then there is no known winning strategy so far. One strategy is that the first player tries to finish the game at the odd number of turns and the second player tries to finish the game at the even number of turns. The player who keeps this strategy will win the game.

CUBE NET

Cube net game is similar to *factorization* game. The difference is to draw the cube net instead of squares. When you do this game, open your eyes wide. There might be a space for a cube net to fit in.

There are 11 cube nets in total (Figure 59). Select one of them and the other player selects one of them as well. Just mark six boxes based on the 11 cube nets. Do it alternately and the person who can no longer play will lose the game.

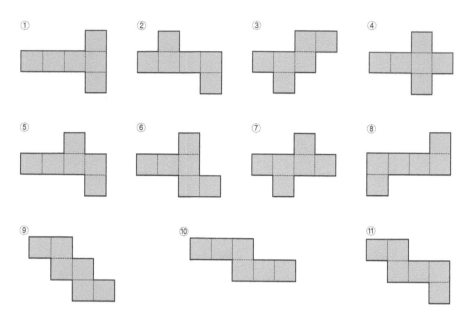

FIGURE 59   11 Cube nets.

In Figure 60, Red plays first and Blue plays second. Suppose that this is Red's turn. Where does Red put? Take a close look at some of six empty boxes in the middle to make a cube net. Therefore, Red can win the game.

In a small board of size 4×4, the first player can draw one of the ten cube nets from Figure 59 since the tenth *cube net* requires five horizontal boxes and wins the game. In a board of size 5×5, Red can always win the game with some simple strategy. For example, Red puts the *T* shape in the center of the board. Then Blue has only two shapes such as ① and ⑩ in Figure 59 which should be placed on the left or right side with respect to *T*. Then Red puts the same shape as Blue in the opposite side. Hence Red wins the game. In the case of board size 6×6, we do not have a winning strategy for either the first player or the second player. The first player should put the last stone at the odd number of turns while the second player should put the last stone at the even number of turns.

ACUTE TRIANGLE

Acute triangle game is a game putting three stones at the same time such that the three stones make an acute triangle. For example, the three red dots form an acute triangle. The three blue dots also form an acute triangle. They play alternately and the person who can no longer put an acute triangle will lose the game

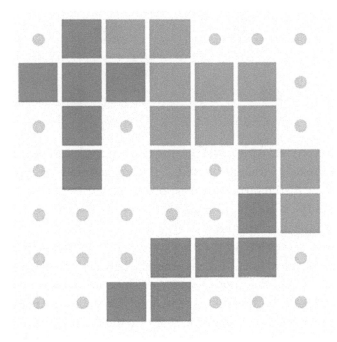

FIGURE 60   A cube net game on the 7×7 board.

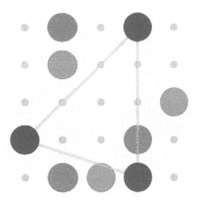

FIGURE 61   Acute triangle game.

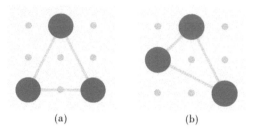

(a)   (b)

For example, on the 3×3 board, there are two types of possible three stones. If we rotate an acute triangle 90, 180, 270 degrees from each type, then we get three more distinct triangles. Therefore, there are eight distinct shapes from (a) and (b). If Red puts one of the types of (a) or (b), Blue can put a different type to win the game.

How about the game on the 3×3 board? It is more complicated, but there are at most five turns. Because there are 16 spots to place stones and three stones are put at the same time to reduce the available spots, there are at most five turns. Therefore, the strategy for Red is to finish the game in the fifth turn and the strategy for Blue is to finish the game in the fourth turn. In general, in the case of the $n \times n$ board for this *acute triangle* game, there are at most $n^2 / 3$ turns. In fact, when you play this game, there are one or two less turns than this maximal number. Therefore, when you make a mistake, the result for winning or losing can change.

As similar games, one can think of *rectangle* games, *obtuse triangle*, and a *general triangle*. Furthermore, an *n-gon game* is possible. Please think about various games and try them.

PROBLEMS

1. (*) When two players play *factorization* game on the 4×4 board, is there a winning strategy for the first player?

2. (**) When two players play *cube net* game on the 6×6 board, find at least five winning strategies for the first player.

3. (**) When two players play *acute triangle* game on the 5×5 board, find at least five winning strategies for the first player? Also, find at least five winning strategies for the second player.

4. (***) Make your own games similar to *factorization* game, *cube net* game, and *acute* game and share them with your friends.

CHAPTER 15

SET Game

Steiner Triple System Game

CHOOSE A SET OF THREE CARDS

SET Game is a real-time game played by one or more people. A genetic engineer Marsha Falco invented this game in 1974 while she was researching genetics of Shepherds. She put symbols in cards to represent information of dogs. Different symbols with different properties represented different traits. She found it fun to explain to other people. Later, she launched Set Enterprises in 1991 to commercialize *SET* and now the company has grown up internationally. This game was selected as one of the top games by Mensa.

SET consists of 81 cards with four types of symbols. Each type has three properties. More precisely, each card contains numbers (1, 2, 3), shapes (diamond, squiggle, oval), shading (solid, striped, open), and color (red, green, purple). Therefore, there are $3 \times 3 \times 3 \times 3 = 81$ cards.

DOI: 10.1201/9781003268024-15

The rule of *SET* is simple. A dealer mixes the 81 cards and displays 12 cards on the table. The goal of the game is to select a set of three cards such that properties are either all same or all different. Such a set of three cards are called a 'Set'. The player who makes a set will take three cards. Then the dealer adds three cards to the remaining nine cards. Play the game until there is no more Set. The player who has the largest number of cards becomes the winner.

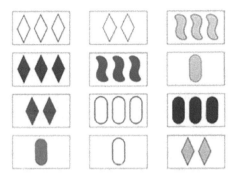

The following three cards form a Set. We need to check whether each type satisfies either of the two conditions, all same or all different. In this case, there are three different numbers, the same shape, the same shading, and the different colors.

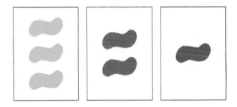

We can display nine cards where each row, each column, and two diagonals form Sets. This is like a magic square in some sense.

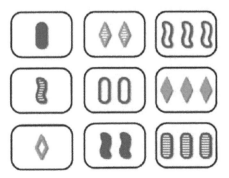

For example, the first column is a Set since all numbers are the same while the shapes, the shadings, and the colors are all different. In fact, all three columns have the same number and the other types are all distinct. However, the three rows have the condition that all properties are different. In each diagonal, only one property is the same and the other three properties are all different.

Although *SET* may not intend any mathematical feature, it can be easily described in terms of a combinational design. One can observe that given any two cards there is a unique third card so that the three cards form a Set. This property was already known in a design theory, known as a balanced incomplete block design (BIBD).

A BIBD is a collection B of b subsets, called blocks, of a finite set X of v elements such that any element of X is contained in the same number r of blocks, every block has the same number k of elements, and each pair of distinct elements appear together in the same number λ of blocks. BIBD are also called as 2-designs, that is, any two elements belong to λ blocks of size k. BIBD are denoted by $2-(v, k, \lambda)$ design. In particular, if $\lambda = 1$ and $k = 3$, such a BIBD is called a Steiner triple system of order v denoted by $S(2, 3, v)$.

For example, when $v = 7$, we have s Steiner triple system of order 7, $S(2, 3, 7)$. This is called the Fano plane. It consists of 7 points and 7 blocks (or lines). Each block contains three points and every pair of points belongs to a unique block.

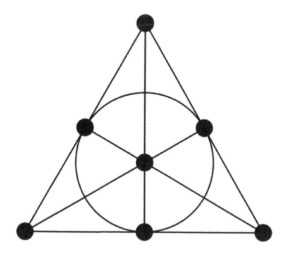

SET has 81 cards or 81 elements. Any pair of two cards belongs to a unique Set since there is a unique third card to form a Set. Each Set corresponds to a block of size 3. Therefore, the 81 cards in *SET* form a Steiner triple system of order 81, that is, *SET* is $S(2, 3, 81)$.

Another interesting question is that how many cards should be placed on the table in order to make sure that one can find at least one set among them. This question makes sense because if there are a few cards on the table, we might not find a set. It might happen that there is no *SET* from the twelve cards on the table. Adding a few more might not guarantee the existence of a *SET*. Here comes the concept of a cap. A *cap* is a collection of cards containing no *SET*. It was proved that the maximum size of a cap is 20. In other words, there is a set of 20 cards where you cannot find a *SET*. The proof of this fact is beyond our scope. But there is a mathematical representation of a cap as follows.

Let $F_3 = \{0, 1, 2\}$ be the field of three elements, where the addition and multiplication are done by modulo 3 operation. These numbers correspond to the three properties such as red, green, purple. We associate each of the four types with one of the three elements from F_3. Therefore, each card in *SET* game corresponds to an element of $F_3^4 = \{(a_1, a_2, a_3, a_4) \mid a_i \in F_3 \text{ for each } i\}$. This accounts for 81 cards in *SET* game.

Let $\mathbf{x}, \mathbf{y}, \mathbf{z}$ be any three elements of F_3^4. One can check that the following conditions are equivalent.

    i. $\mathbf{x}, \mathbf{y},$ and $\mathbf{z}$ form a *SET*,

    ii. $\mathbf{x} + \mathbf{y} + \mathbf{z} = \mathbf{0}$,

    iii. $\mathbf{x}, \mathbf{y},$ and $\mathbf{z}$ are in a line.

For example, let $\mathbf{x} = (0, 1, 2, 0)$, $\mathbf{y} = (1, 1, 2, 0)$. Then to make a *SET*, we need $\mathbf{z} = (2, 1, 2, 0)$ satisfying $\mathbf{x} + \mathbf{y} + \mathbf{z} = \mathbf{0}$ so that each coordinate satisfies $0 + 1 + 2 = 0 \pmod 3$ or $1 + 1 + 1 = 0 \pmod 3$, $2 + 2 + 2 = 0 \pmod 3$, or $0 + 0 + 0 = 0 \pmod 3$.

Therefore, a cap in F_3^4 is a set of elements in F_3^4 such that no three elements (points) are not in a line. As discussed above, the maximum size of a cap in F_3^4 is 20. Instead of proving this, we consider a cap in F_3^2 as follows.

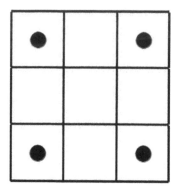

There are no three points in a straight line, where we allow the side points are connected with the other side points just like a torus. One can check that the maximum size of a cap in F_3^2 is 4. In general, let $\alpha(n)$ be the maximum size of a cap in F_3^n. Its exact values are known from https://oeis.org/A090245 up to $n = 6$, as shown below.

n	1	2	3	4	5	6	7
$\alpha(n)$	2	4	9	20	45	112	unknown

Therefore, it will be a challenging open problem to compute $\alpha(7)$.

PROBLEMS

1. Find at least three Sets in the following. In fact, there are six Sets.

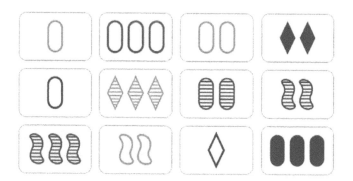

(selected from https://www.setgame.com/set/puzzle/yesterday February 2, 2023)

One example of a Set is

2. (*) Find another cap of size 4 in F_3^2, different from the above example.

3. (*) Show that the maximum size of a cap in F_3^2 is 4.

4. (**) Show that the maximum size of a cap in F_3^3 is 9.

Answer for Problem 1

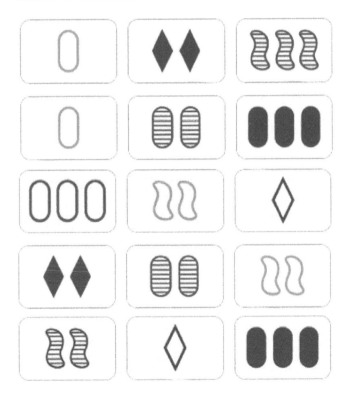

CHAPTER 16

Dobble Game

Finite Projective Plane Game

ANY TWO CARDS SHARE EXACTLY ONE SYMBOL

Dobble is a speedy game played by two or more people. It is also known as *Spot It* in the USA. Any two cards of *Dobble* share exactly one symbol.

It goes back to Jacques Cottereau who introduced a set of 31 cards in 1976. Each card has six images of insects and any two cards share exactly one image. We note that $31 = 5^2 + 5 + 1$. Here, the number $n = 5$ plays an important parameter. If $n = 2$, then there will be seven cards such that

DOI: 10.1201/9781003268024-16

each card has three symbols and any two cards share exactly one symbol. These seven cards are shown below.

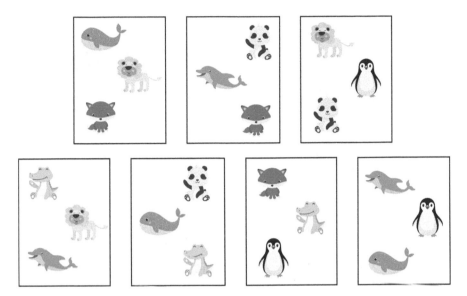

In 2008, a game designer Denis Blanchot added more cards to develop the idea to create the current *Dobble* which has been commercialized by the company Asmodee. *Dobble* has 55 cards, each of which has eight symbols. It can be played by two to five players. The rule of the game is easy to follow.

1. Place all the cards in the center of the table with face up.

2. Each player takes one card with face up.

3. When the game starts, each player shouts a symbol that is common with his or her top card. If this is correct, that player takes the card and puts it on the player's top card.

4. The game continues until all the cards on the table are exhausted completely.

5. Rank the players according to the number of cards they collected.

There are other rules of the game. All the rules are based on the special property that any two cards share exactly one symbol.

In the following two cards, what is a common symbol?

The following symbol is the only common symbol.

FINITE PROJECTIVE PLANE OF ORDER N

It is rather mysterious why any two cards from the 55 cards share a symbol and it is the only one. To explain this, we introduce the concept of a finite projective plane of order n. Here is the definition.

A projective plane of order n ($n \geq 2$) is a finite set of points and lines (defined as sets of points) satisfying the following conditions:

1. Every line contains $n+1$ points.

2. Every point lies on $n+1$ lines.

3. Any two distinct lines intersect in a unique point.

4. Any two distinct points lie on a unique line.

When $n = 2$, the Fano plane is the unique projective plane of order 2 shown below.

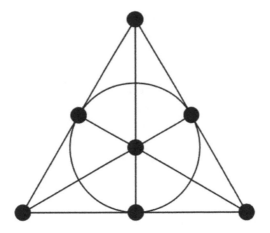

From the previous chapter, it is also known as a Steiner triple system of order 7 denoted by $S(2, 3, 7)$. Each line of the Fano plane consists of $3 = 2 + 1$ points and corresponds to a block of size 3. Each point lies on the three lines. Any two distinct lines meet at a unique point. Any two distinct points determine a unique line. Therefore, the Fano plane is a finite projective plane of order 2. In terms of a BIBD in the previous chapter, a projective plane of order n is the same as $2 - (n^2 + n + 1, n + 1, 1)$ design. This means that a projective plane of order n has exactly $n^2 + n + 1$ points and $n^2 + n + 1$ lines.

Now, this explains why Cottereau considered 31 cards with six images of insects. He was thinking of a projective plane of order 5 so that each card (corresponding to a line) consists of $5 + 1 = 6$ images and there are $5^2 + 5 + 1 = 31$ cards together. Now we introduce a theorem telling the existence of a projective plane of order n.

Theorem. If $n = p^k$, where p is a prime and k is a positive integer, then there exists a projective plane of order n.

Interestingly, the orders of all the known finite projective planes are a prime power. Therefore, finding a finite projective plane of other orders has been an active open problem. One necessary condition for the existence of a finite projective plane of order n was known as the Bruck-Ryser theorem.

Theorem (Bruck-Ryser). *If there is a finite projective plane of order n and n is congruent to 1 or 2 (mod 4), then n is the sum of two squares.*

The first non-prime power order is 6. Since 6 is congruent to 2 (mod 4), 6 must be the sum of two squares if there exists a finite projective plane of order 6. It is easy to see that this is not possible. Therefore, we know that there is no finite projective plane of order 6.

We note that the converse of the Bruck-Ryser theorem does not hold in general. For example, when $n = 10$, 10 is the sum of 1 and 9. Since 1970, there was a hope that there would exist a projective plane of order 10. Using the knowledge of the weight distribution of the code generated by the 111 by 111 incidence matrix of a putative projective plane of order 10, Lam, Thiel, and Swiercz in 1989 proved that there does not exist a finite projective plane of order 10.

Now, we can see that *Dobble* game is based on a finite projective plane of order 7. There should be $7^2 + 7 + 1 = 57$ cards. We recall that *Dobble* has only 55 cards. The company removed two cards. The author has calculated that the two missing cards are as follows.

PROBLEMS

1. (*) In the above two missing cards, what is a common symbol?

2. (**) Check that the following figure represents a finite projective plane of order 3. There should be 13 points and 13 lines. Identify all 13 lines.

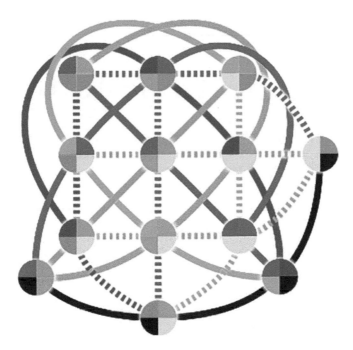

3. (*) The following is called an affine plane of order 3. Each line contains exactly three points. How many lines are there? What is the difference and relation between an affine plane of order 3 and a projective plane of order 3?

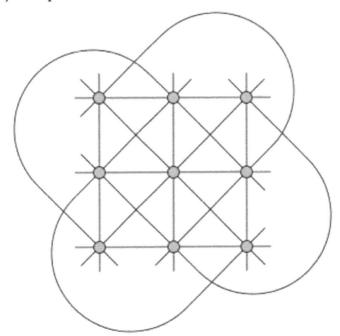

CHAPTER 17

Find-a-Best-Friend Game

A Game of the Perfect Hamming Code

HOW TO PLAY THE GAME

In this chapter, we introduce a game called *"Find-a-best-friend"* invented by the author in 2018 who got the motivation from error-correcting codes.

Find-a-best-friend is a two or more players' game with the goal that each player should find a pair of cards (called the matching cards) whose colors differ in exactly one place.

DOI: 10.1201/9781003268024-17

Find-a-best-friend consists of 48 cards. Twelve cards have a grey color background, called the Hamming cards. The other 36 cards have a white color background, called the best-friend cards.

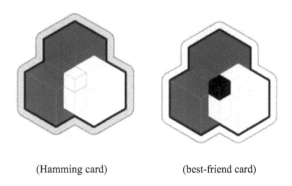

(Hamming card)   (best-friend card)

One Hamming card and one best-friend card are called a "match" if there is only one color difference between them. See below. Only yellow color is different.

The following two cards are not matching cards since there are three colors difference.

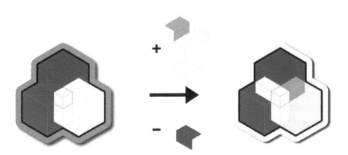

Below is the order of the game.

1. Determine who will play first.

2. Mix the Hamming cards and best-friend cards separately.

3. Place the 12 Hamming cards in the center on the table with face down.

4. Each player is given five best-friend cards and put them face up.

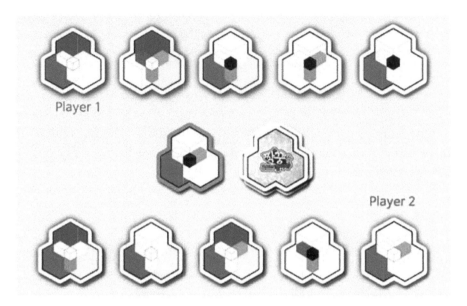

5. The first player picks the Hamming card in the center and face up.

6. Each player checks whether each player has a best-friend card which is a best-friend with the picked Hamming card as shown below. If so, shout "match".

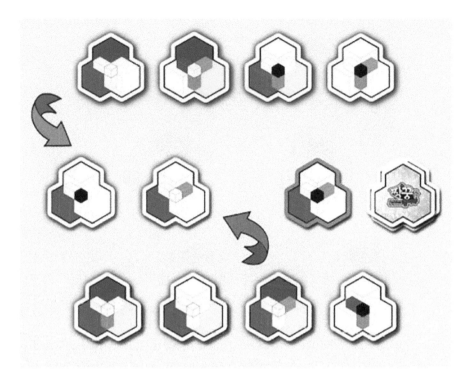

7. If there is no more best-friend card to put down, then move to a next player to turn up the Hamming card.

8. The next player puts the Hamming card on the previous Hamming card.

9. Players repeat this process until all the Hamming cards are exhausted.

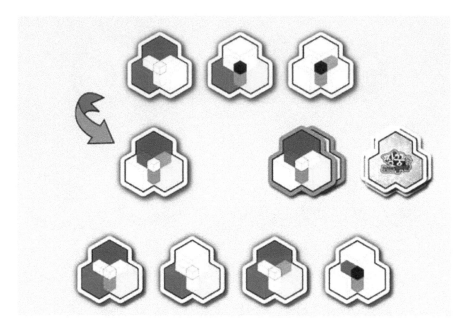

10. One can use "chance" only one time if there is exactly two colors difference between the Hamming card and one's own best-friend card.

→ Chance

11. (penalty rule)

   If a player puts down a wrong best-friend card to match with a Hamming card or the chance call was wrong, the player should get an extra best-friend card as a penalty.

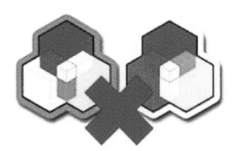

If a player does not say "match", "pass", or "chance" within the agreed time limit such as 5 or 10 seconds, then the player should get an extra best-friend card as a penalty.

12. (rank) The player who first matches all the five best-friend cards with the Hamming cards will be the winner. So, the rank is based on the order of the players finishing the game. When there is a tie, the player who shouts "match" first will be ranked higher.

MATHEMATICS OF *FIND-A-BEST-FRIEND*

Now we can describe the mathematics behind the game. We remark that *Find-a-best-friend* is based on the Hamming code of length 7.

The Hamming code of length 7 is the set of binary sequences of length 7 satisfying a certain condition so that it can correct any one-error out of the seven positions. The Hamming code and its generalization were found by Richard Hamming in the late 1940s. He shared his idea on the existence of one-error-correcting code with Claude Shannon. In 1948, Shannon published a seminal paper on Information Theory titled as "A mathematical theory of communication" which opened the area of Coding Theory or the Theory of Error-Correcting Codes.

One of the ways to represent the Hamming code of length 7 is to combine three circles like a Venn Diagram.

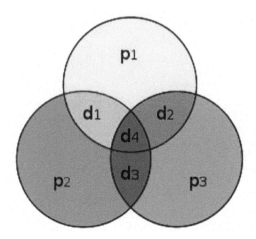

There are seven colored regions. Each region is placed a 0 or 1. The four intersection regions d_1, d_2, d_3, d_4 can take any values of 0 or 1, meaning that these are independent variables. Hence, there are 16 possible values of d_1, d_2, d_3, d_4. The other three variables p_1, p_2, p_3 are dependent. More precisely,

i. $p_1 = d_1 + d_2 + d_4$ (mod 2)

ii. $p_2 = d_1 + d_3 + d_4$ (mod 2)

iii. $p_3 = d_2 + d_3 + d_4$ (mod 2)

For example, if $d_1 = 1$, $d_2 = 0$, $d_3 = 0$, $d_4 = 0$, then $p_1 = 1$, $p_2 = 1$, $p_3 = 0$. Then, we obtain a vector $(d_1, d_2, d_3, d_4, p_1, p_2, p_3) = (1, 0, 0, 0, 1, 1, 0)$, called a codeword of the Hamming code. We can rewrite i, ii, iii as the parity-check equations as follows.

iv. $d_1 + d_2 + d_4 + p_1 = 0$ (mod 2)

v. $d_1 + d_3 + d_4 + p_2 = 0$ (mod 2)

vi. $d_2 + d_3 + d_4 + p_3 = 0$ (mod 2)

Now, we can visualize iv, v, and vi by the condition that each circle has an even number of ones including no ones. If each circle has only zeros, it corresponds to the zero vector $0 = (0, 0, 0, 0, 0, 0, 0)$, which satisfies the three conditions iv, v, and vi.

In fact, these conditions may help to correct any one-error out of the seven positions. For example,

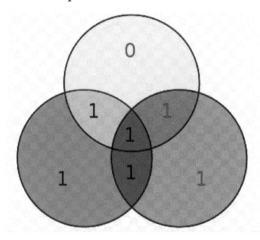

The top circle has three 1 s. Therefore, it does not satisfy the condition iv. However, the left circle has four 1's satisfying the condition v, and the right circle has four 1's satisfying the condition vi. It is enough to modify the first circle. In particular, if the zero in the first circle is changed to 1, then the first circle has four 1's, which satisfies the condition iv. This change does not affect the parity of the other two circles. Therefore, we have corrected one error.

Now we can suggest a decoding algorithm as follows.

1. Check if all circles have even number of ones. If not, go to step 2.

2. If only one circle has an odd number of ones, then switch the outmost bit, which does not overlap with other circles. Then exit.

3. If exactly two circles have odd numbers of ones, then switch the bit in the common circle but not in the center region. Then exit.

4. If the three circles have odd numbers of ones, then switch the bit in the center region. Then exit.

After this algorithm, we will have a correct Hamming codeword.

Therefore, *Find-a-best-friend* game uses the Venn diagram representation of the Hamming [7,4] code. Instead of the circles, the author designed three cubes to make them into a 3-dimensional figure.

As one can see, the Hamming cards have an even number of colors for each cube and the best-friend cards have an odd number of colors for some cube. The Hamming cards correspond to the Hamming codewords. Therefore, the process that the best-friend card is matched with the Hamming card using the help of colors is the same as decoding a vector of length 7 with one error to a Hamming codeword.

PROBLEMS

1. (**) *Find-a-best-friend* cards have 12 Hamming cards. It is a little strange since there are 16 Hamming codewords. Those 12 Hamming cards have only one axis of symmetry. The Hamming card below has one axis of symmetry which is the line passing the purple block and the white cube.

Describe the other 11 Hamming cards.

Find the four cards which do not satisfy this condition but correspond to the four Hamming codewords.

2. (**) Using the property of (1), describe why there are the 36 best-friend cards which match with the 12 Hamming cards. How many best-friend cards are there matching with a given Hamming card?

3. (*) Using the relation on the Hamming code of length 7, find all codewords of length 7 of the Hamming code.

4. (*) Prove that the decoding algorithm for the Hamming [7,4] code described in this chapter really works.

5. (*) Enjoy this game with another rule described as follows.

   (1) Place twelve Hamming cards as follows.

   (2) Share best-friend cards equally. For example, 18 cards for two players, 12 cards for three players, etc.

   (3) When it starts, every player matches his/her best-friend cards with the faced-up Hamming cards as quickly as possible.

   (4) Rank the players based on the order of completing the matching.

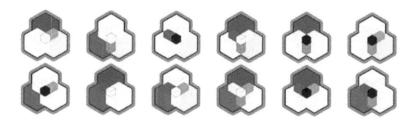

9 781032 213057